Overview of Blotting Techniques

Overview of Blotting Techniques

Edited by:

Shiv Sanjeevi Sripathi
and Prerna Pandey

Delve
PUBLISHING

www.delvepublishing.com

Overview of Blotting Techniques

Shiv Sanjeevi Sripathi and Prerna Pandey

Delve Publishing
224 Shoreacres Road
Burlington, ON L7L 2H2
Canada
www.delvepublishing.com

Email: orders@arclereducation.com

ISBN: 978-1-77407-695-8 (Hardcover)

DECLARATION

ABOUT THE EDITORS

ShivSanjeevi Sripathi completed his Masters in Biotechnology from Mumbai University in 2008. He was awarded for academic excellence in both his Bachelors and Masters for securing second rank in Mumbai University in 2006 and first rank in his college: Kishinchand chellaram College. For his Masters he secured first rank in his college KET's V.G.Vaze College. He qualified CSIR and NET and and TOEFL in September 2008. He then worked on a stem cell project at the Specialized Centre for Cell Based Therapy (SCCT),KEM Hospital at Mumbai on a project entitled,: Isolation & detection of stem cells from Human Umbilical cord/amniotic membrane" following which he worked at Junior Research Fellow at Microbiology & Cell Biology Department, Indian Institute of Science, Bangalore from August 2009 to November 2011. The field of work involved cloning of cell wall genes and transcription factors in *E.coli & M.smegmatis*. As a writer, he has authored and co-authored 25 books with Delve and Arcler International Press. He loves to read and share on interesting aspects of life sciences in books. In his free time he loves to travel and explore ancient customs and traditions.

Dr Prerna Pandey, a biotechnologist with several years of wet lab research experience, has worked as a senior research fellow at International Center for Genetic Engineering and Biotechnology, New Delhi. Her field of research in PhD included isolation and molecular characterization of Geminiviruses, genome sequencing, gene annotation, as well as gene silencing using the RNA interference technology. She has published papers in peer-reviewed journals in the field and has a submitted a number of annotated Geminiviral genome sequences in the GenBank, including two novel ones. Prerna has also completed her editing and proof-reading courses from SfEP, UK and now works as a freelance scientific editor, writer and a subject matter expert. She has edited and co-authored several books on a range of topics in Life-Sciences. When Prerna is not busy with her assignments, she loves visiting historical places.

TABLE OF CONTENTS

LIST OF CONTRIBUTORS

Shivsanjeevi Sripathi

Prerna Pandey

Trent A. J. Butler
Mothers and Babies Research Centre, School of Medicine and Public Health, Faculty of Health and Medicine, University of Newcastle, Callaghan, NSW, 2308, Australia
Hunter Medical Research Institute, 1 Kookaburra Circuit, New Lambton Heights, NSW, 2305, Australia

Jonathan W. Paul
Mothers and Babies Research Centre, School of Medicine and Public Health, Faculty of Health and Medicine, University of Newcastle, Callaghan, NSW, 2308, Australia
Hunter Medical Research Institute, 1 Kookaburra Circuit, New Lambton Heights, NSW, 2305, Australia

Eng-Cheng Chan
Mothers and Babies Research Centre, School of Medicine and Public Health, Faculty of Health and Medicine, University of Newcastle, Callaghan, NSW, 2308, Australia
Hunter Medical Research Institute, 1 Kookaburra Circuit, New Lambton Heights, NSW, 2305, Australia

Roger Smith
Mothers and Babies Research Centre, School of Medicine and Public Health, Faculty of Health and Medicine, University of Newcastle, Callaghan, NSW, 2308, Australia
Hunter Medical Research Institute, 1 Kookaburra Circuit, New Lambton Heights, NSW, 2305, Australia
John Hunter Hospital, New Lambton Heights, NSW, 2305, Australia

Jorge M. Tolosa
Mothers and Babies Research Centre, School of Medicine and Public Health, Faculty of Health and Medicine, University of Newcastle, Callaghan, NSW, 2308, Australia
Hunter Medical Research Institute, 1 Kookaburra Circuit, New Lambton Heights, NSW, 2305, Australia

Hiroyuki Tanaka
Graduate School of Pharmaceutical Science, Kyushu University, 3-1-1 Maidashi, Higashi-ku, Fukuoka 812-0855, Japan

Osamu Morinaga
Faculty of Pharmaceutical Science, Nagasaki International University, 2827-7 Huis Ten Bosch, Sasebo 859-3298, Japan

Takuhiro Uto
Faculty of Pharmaceutical Science, Nagasaki International University, 2827-7 Huis Ten Bosch, Sasebo 859-3298, Japan

Shunsuke Fuji
Faculty of Health Management, Nagasaki International University, 2827-7 Huis Ten Bosch, Sasebo 859-3298, Japan

Frederick Asare Aboagye
Phytochemistry Department, Centre for Scientific Research into Plant Medicine, University of Ghana, MampongAkuapem, Ghana

Nguyen Huu Tung
Faculty of Pharmaceutical Science, Nagasaki International University, 2827-7 Huis Ten Bosch, Sasebo 859-3298, Japan

Xiao Wei Li
State Key Laboratory of Natural and Biomimetic Drugs, School of Pharmaceutical Sciences, Peking University, No.38 Xue-yuan Road, Haidian District, Beijing 100191, China

Waraporn Putalun
Faculty of Pharmaceutical Science, Khon Kaen University, Khon Kaen 40002, Thailand

Yukihiro Shoyama
Faculty of Pharmaceutical Science, Nagasaki International University, 2827-7 Huis Ten Bosch, Sasebo 859-3298, Japan

Andrea Degasperi
Systems Biology Ireland, University College Dublin, Dublin, Republic of Ireland

Marc R. Birtwistle
Department of Pharmacology and Systems Therapeutics, Icahn School of Medicine at Mount Sinai, New York, New York, United States of America

Natalia Volinsky
Systems Biology Ireland, University College Dublin, Dublin, Republic of Ireland

Jens Rauch
Systems Biology Ireland, University College Dublin, Dublin, Republic of Ireland

Walter Kolch
Systems Biology Ireland, University College Dublin, Dublin, Republic of Ireland
Conway Institute of Biomolecular & Biomedical Research, University College Dublin, Dublin, Republic of Ireland
School of Medicine and Medical Science, University College Dublin, Dublin, Republic of Ireland

Boris N. Kholodenko
Systems Biology Ireland, University College Dublin, Dublin, Republic of Ireland
Conway Institute of Biomolecular & Biomedical Research, University College Dublin, Dublin, Republic of Ireland
School of Medicine and Medical Science, University College Dublin, Dublin, Republic of Ireland

Sean C. Taylor
Bio-Rad Laboratories, 1329 Meyerside Drive, Mississauga, ON, Canada

Anton Posch
Bio-Rad Laboratories GmbH, Heidemann Street 164, 80939 Munich, Germany

Brad L. Ericson
Biology Department, University of Nebraska at Kearney, 2401 11th Avenue, Kearney, NE 68849, USA

Darby J. Carlson
Biology Department, University of Nebraska at Kearney, 2401 11th Avenue, Kearney, NE 68849, USA

Kimberly A. Carlson
Biology Department, University of Nebraska at Kearney, 2401 11th Avenue, Kearney, NE 68849, USA

Yukari Kondo
Division of Transcriptomics, Medical Institute of Bioregulation, Kyushu University, Fukuoka, Japan

Shinichiro Higa
Division of Transcriptomics, Medical Institute of Bioregulation, Kyushu University, Fukuoka, Japan
Department of Cellular Biochemistry, Graduate School of Pharmaceutical Sciences, Kyushu University, Fukuoka, Japan

Takeshi Iwasaki
Division of Transcriptomics, Medical Institute of Bioregulation, Kyushu University, Fukuoka, Japan

Tomoya Matsumoto
Sales Department, Microscope Division, Keyence, Osaka, Japan

Kazumitsu Maehara
Division of Transcriptomics, Medical Institute of Bioregulation, Kyushu University, Fukuoka, Japan

Akihito Harada
Division of Transcriptomics, Medical Institute of Bioregulation, Kyushu University, Fukuoka, Japan

Yoshihiro Baba
Division of Immunology and Genome Biology, Medical Institute of Bioregulation, Kyushu University, Fukuoka, Japan

Masatoshi Fujita
Department of Cellular Biochemistry, Graduate School of Pharmaceutical Sciences, Kyushu University, Fukuoka, Japan

Yasuyuki Ohkawa
Division of Transcriptomics, Medical Institute of Bioregulation, Kyushu University, Fukuoka, Japan

Samantha L. Eaton
Division of Neurobiology, The Roslin Institute and Royal (Dick) School of Veterinary Studies, University of Edinburgh, Edinburgh, United Kingdom

Sarah L. Roche
Centre for Integrative Physiology, University of Edinburgh, Edinburgh, United Kingdom

Maica Llavero Hurtado
Division of Neurobiology, The Roslin Institute and Royal (Dick) School of Veterinary Studies, University of Edinburgh, Edinburgh, United Kingdom

Karla J. Oldknow
Developmental Biology, The Roslin Institute and Royal (Dick) School of Veterinary Studies, University of Edinburgh, Edinburgh, United Kingdom

Colin Farquharson
Developmental Biology, The Roslin Institute and Royal (Dick) School of Veterinary Studies, University of Edinburgh, Edinburgh, United Kingdom

Thomas H. Gillingwater
Centre for Integrative Physiology, University of Edinburgh, Edinburgh, United Kingdom
Euan MacDonald Centre for Motor Neurone Disease Research, University of Edinburgh, Edinburgh, United Kingdom

Thomas M. Wishart
Division of Neurobiology, The Roslin Institute and Royal (Dick) School of Veterinary Studies, University of Edinburgh, Edinburgh, United Kingdom
Euan MacDonald Centre for Motor Neurone Disease Research, University of Edinburgh, Edinburgh, United Kingdom

LIST OF ABBREVIATIONS

AFU	Arbitrary Fluorescence Units
BPB	Bromophenol Blue
BSA	Bovine Serum Albumin
CCD	Charge Coupled Device
CD	Circular Dichroism
CE	Capillary Electrophoresis
CGE	Capillary Gel Electrophoresis
CNS	Central Nervous System
CV	Coefficient of Variation
DMEM	Dulbecco's modified Eagle's medium
ECL	Enhanced Chemiluminescence
FD	Fold Difference
GAPDH	Glyceraldehyde -Phosphate Dehydrogenase
GC	Glycyrrhizin
HKPs	Housekeeping Proteins
HSCCC	High-Speed Counter Current Chromatography
IC	Ion Chromatography
IMAC	Immobilized Metal Ion Affinity Chromatography
IRES	Internal Ribosomal Entry site
LC	Loading Control
LDS	Lithium Dodecyl Sulphate
MAb	Monoclonal Antibody
MALDI	Matrix-assisted laser desorption/ionization
NMR	Nuclear Magnetic Resonance
O.D.	Optical Density
OBB	Odyssey blocking buffer
ORFs	Open Reading Frames
PBS	Phosphate Buffered Saline

PDVF	Polyvinylidene Fluoride
PMT	Photomultiplier Tube
PTM	Post-Translational Modifications
QWB	Quantitative Western Blotting
SMA	Spinal Muscular Atrophy
SPR	Surface plasmon resonance
TPA	Total Protein Analysis
WB	Western blotting
WGM	Whispering Gallery Mode

PREFACE

Following the commonly used techniques like electrophoresis, separated nucleic acids and proteins can be transferred to a membrane where they are analyzed or detected. Depending on the molecule being analyzed, different blotting methods like Southern blot, Northern blot, Western blot and Eastern blotting are applied.

Given the importance of blotting in the lab, this book comes at an opportune time. The book discusses the protocols and applications of blotting-Southern, western, northern and eastern blotting. Case studies have also presented to help the reader gain insights of the practicality and applicability of blotting in today's world. Each chapter has been presented with images using simple language so that even a reader from a non-scientific background can easily grasp the concepts. The essential protocols have been presented with a section devoted to troubleshooting. A researcher can easily check for possible causes and solutions at one instant.

This is one book that provides concise and vital information of all the important types of blotting along with relevant references that can offer insights to further reading. The editors hope that this book will serve as a lab staple just like the aforementioned blotting techniques.

Editors

1

Western Blotting-the Works

Shivsanjeevi Sripathi

ShivSanjeevi Sripathi completed his Masters in Biotechnology from Mumbai University in 2008. He was awarded for academic excellence in both his Bachelors and Masters for securing second rank in Mumbai University in 2006 and first rank in his college: Kishinchand chellaram College. For his Masters he secured first rank in his college KET's V.G.Vaze College. He qualified CSIR and NET and and TOEFL in September 2008. He then worked on a stem cell project at the Specialized Centre for Cell Based Therapy (SCCT),KEM Hospital at Mumbai on a project entitled, : Isolation & detection of stem cells from Human Umbilical cord/ amniotic membrane" following which he worked at Junior Research Fellow at Microbiology & Cell Biology Department, Indian Institute of Science, Bangalore from August 2009 to November 2011. The field of work involved cloning of cell wall genes and transcription factors in *E.coli & M.smegmatis*. As a writer, he has authored and co-authored 25 books with Delve and Arcler International Press. He loves to read and share on interesting aspects of life sciences in books. In his free time he loves to travel and explore ancient customs and traditions.

1.1 CONCEPT

The technique called Western blotting (WB) is described as semi-quantitative to analyze proteins that made its entry in 1979 and has become a lab-staple. The approach is described as orthogonal: separation of proteins on the basis of size initially using gel electrophoresis followed by the use of antibodies specific for the protein of interest. Indeed, antibodies have assumed a central role in diagnosis, biotechnology and research given their ability to bind specific proteins (Sanders et al, 2016).

Western blot finds application in the separation and identification of proteins. The process involves the resolution of proteins through gel electrophoresis. A band of each protein is transferred to a membrane that is treated with labeled antibodies specific to the protein under study. To retain only the antibodies that have bound the protein, washing of unbound antibodies is done. The film is then developed based on the label used to reveal a band where the protein of interest is present. As the quantity of protein present relates to the thickness of the band formed, the amount of the test protein can be determined by running a standard.

1.2 THE DETAILS

1.2.1 Sample preparation

Frequently used samples for running western blots are cell lysates. The process of protein extraction looks at isolating all proteins present in the cytoplasm. The presence of mechanical barriers in tissues warrants the use of mechanical means to break the tissues such as sonication and homogenization. The use of cold temperatures with protease inhibitors avoids protein denaturation.

This is followed by measuring the concentration of proteins using a spectrophotometer. This allows the comparison of the samples being loaded on an equivalent basis. The mass and volume can also be determined from the concentration. The samples are appropriately diluted in a loading buffer followed by heating to denature the folding but retaining the disulfide bonds and also prevent the charge neutralization of amino acids. The buffer also contains glycerol to allow the samples to sink easily into the wells of the gel and a tracking dye (bromophenol blue) to track the separation.

The presence of appropriate controls is mandatory; a positive control is a known source of the protein under study such as the purified protein or a control lysate is used. In order to ensure that the staining is specific, a negative control like β-actin is utilized.

1.2.2 Gel electrophoresis

There are 2 gels used: stacking and separating gel. The stacking gel is porous with a lower acrylamide concentration and acidic pH of 6.8 that allows the formation of thin, sharply defined bands. The gel on the bottom known as the resolving or separating gel contains higher polyacrylamide that allows the formation of narrow-pore gels. This allows the separation of protein samples with higher resolution as smaller proteins travel faster than larger proteins.

The preparation of gels involves the pouring of the gels between two glass or plastic plates. As the proteins carry a negative charge, the application of voltage allows them to travel to the positive electrode. Following the loading of samples and marker on the wells, running of the gel is done-care should be taken to maintain the optimum voltage as higher voltages can cause overheating and distortion of bands.

1.2.3 Blotting

The separated protein bands are transferred to a membrane. This transfer involves the application of an electric field oriented perpendicular to the surface of the gel. This causes the proteins to be transferred onto the membrane.

This involves the formation of a sandwich made of the membrane that is placed between the gel surface and the positive electrode. There is also a fiber pad (sponge) at each end, and filter papers to protect the gel and blotting membrane.

The membrane being a vital part of the transfer can be nitrocellulose and PVDF. The former displays a high affinity to proteins and cannot be reprobed. The PVDF membranes can allow reprobing, storage and better mechanical support. Care should be taken to wash the membrane carefully to lower the background.

Figure 1.1 below shows the arrangement below:

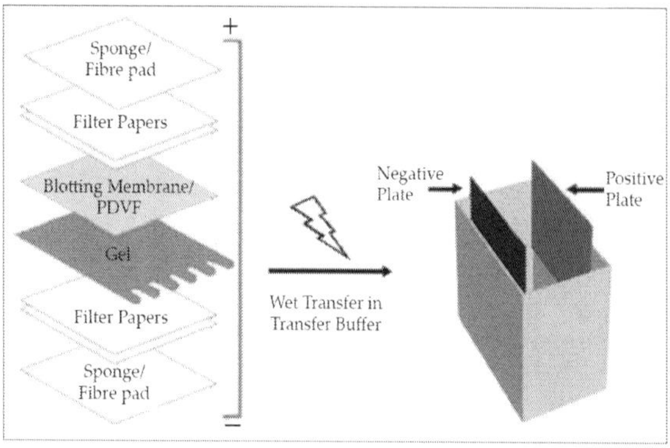

The development of a clear image requires close contact of gel and membrane to ensure a clear image. In order to allow the migration of the proteins from the gel to the membrane, the membrane should be placed between the gel and the positive electrode. This electrophoretic transfer can either be semi-dry or wet.

1.2.4 Washing, blocking and incubation of antibody

This step prevents what is termed as non-specific binding of antibodies to the membrane. Blocking usually involves the use of 5% BSA or nonfat dried milk diluted in TBST to lower the background. The use of nonfat dried milk is not expensive and is also easily available. Care must be taken to check the compatibility of this blocking agent with detection labels. As milk contains casein (a phosphoprotein) and biotin, this blocking agent cannot be used for biotin and AP antibody labels. In this case, BSA is preferred to lower the interference.

The concentration of the antibody used in an assay is based on the instruction by the manufacturer. The process of washing becomes important here to prevent excess background due to unbound antibody. Care should be taken to ensure the washing However, the membrane should not be left to wash for a long time, as it can also reduce the signal. The detection involves the label attached to the antibody.

1.2.5 Quantification

As mentioned earlier, data from a western blot is termed semi-quantitative. This is on account of comparing the levels of proteins and not measuring the quantity absolutely. This is due to differences in the loading and transfer rates between the samples in separate lanes that require precise standardization. Additionally, there is no linearity in the signals across the concentration range of samples.

1.3 GENERAL PROTOCOL FOR WESTERN BLOTTING

1.3.1 Cell lysis

For the case of adherent cells, the protocol for extracting proteins is below (Mahmood and Yang, 2012)

- The cells are washed in the tissue culture flask or dish placed on ice by adding cold phosphate buffered saline (PBS) and rocking gently. Discard the PBS.
- Add PBS and use a cell scraper to dislodge the cells. Pipette the mixture into microcentrifuge tubes.
- Centrifuge at 1500 rpm for 5 minutes and discard the supernatant.
- 180 μL of ice-cold cell lysis buffer with 20 μL fresh protease inhibitor cocktail is added. Incubation is done for 30 minutes on ice, followed by centrifugation for 10 minutes at 12,000 rpm at 4°C.
- The supernatant is transferred to a fresh tube and stored on ice or frozen at -20°C or -80°C.
- The concentration of the protein is quantified using a spectrophotometer.

Tip: If protein concentration is not enough after extraction, it is advised to repeat the procedure with a higher proportion of the protease inhibitor cocktail.

1.3.2 Preparing the sample

- The volume of the protein is determined to ensure 50 μg in each well. This is calculated using the expression:

Concentration=Mass/Volume

- Add 5 μL sample buffer to the sample, and make the volume in each lane equalized using double-distilled H_2O (dd H_2O). Mix well. (Tip: Recommended total volume is15 μL per lane).

- The samples are heated using a dry plate at 100°C for 5 minutes.

1.3.3 Preparing the gels

1. 10% stacking gel solution is prepared followed by assembling the rack for gel solidification as shown below in Figure 1.2 (Mahmood and Yang, 2012):

2. The stacking gel solution is carefully added until the level is at par with the green bar holding the glass plates. Add water till the top. Wait until the gel solidifies. (Tip: The use of a suction pipette can make the process of gel addition to the glass plate easier). Refer Figure 1.3 below (Mahmood and Yang, 2012):

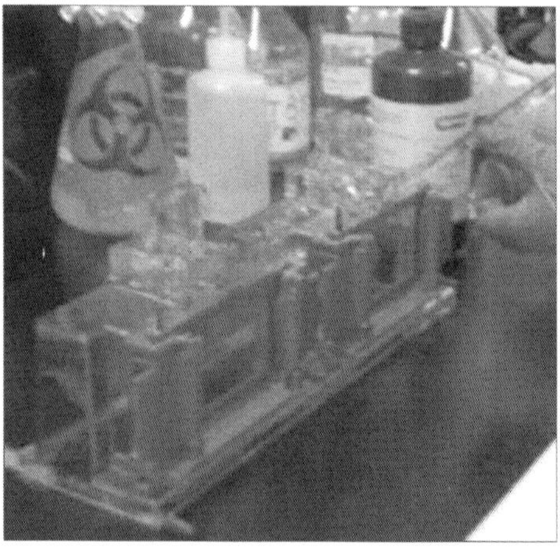

3. The stacking gel is overlaid with the separating gel, after removing the water. (Tip: The entire apparatus can be tilted and use a paper towel to remove the water).

4. Insert the comb, making sure that there are no air bubbles.
5. Wait until the gel is solidified. (Tip: To check if solidification is complete leave some gel solution in another tube and check its progress).

Table 1.1 below shows the composition of the gels (Mahmood and Yang, 2012):

10% Stacking gel	dd H_2O	3 mL
	1 M Tris-HCl	2.1 mL (pH 8.9)
	30% Acr Bis	2.8 mL
	10%SDS	80 μL
	10%APS*	56 μL
	TEMED*	6 μL
6% Separating gel	dd H_2O	2 mL
	1M Tris-HCl	400 μL (pH 6.7)
	30% Acr Bis	600 μL
	10%SDS*	36 μL
	10%APS*	24 μL
	TEMED	4 μL

*, APS: Ammonium Persulfate ; TEMED: Tetramethylethylenediamine; SDS: Sodium dodecyl sulfate

1.3.4 Electrophoresis

The running buffer is poured into the electrophorator as shown in Figure 1.4 below (Mahmood and Yang, 2012):

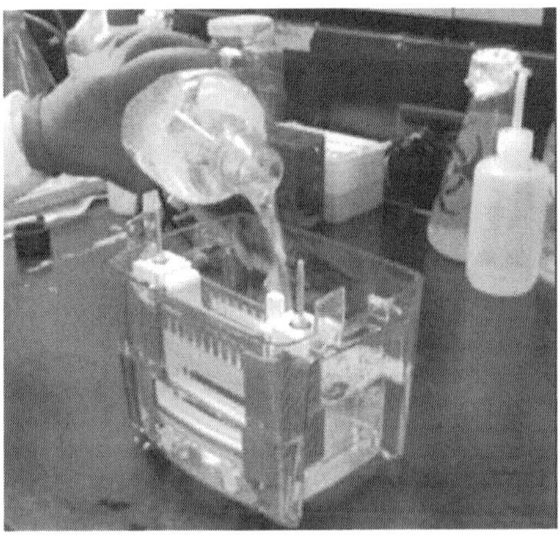

2. The gel is placed inside the electrophorator and connected to a power supply. (Tip: When connecting to the power source always connect red to red, and black to black).

3. Ensure that the buffer covers the gel completely, and remove the comb carefully.
4. The marker (6 μL) followed by samples (15 μL) are loaded in the wells as shown in Figure 1.5 below: (Mahmood and Yang, 2012)

5. The gel is run at low voltage (60 V) for the separating gel; 140 V for the stacking gel [Figure 1.6 below (left;a) represents samples running through the stacking gel at a lower voltage) (Right: b): Samples running through the separating gel at higher voltage) (Mahmood and Yang, 2012):

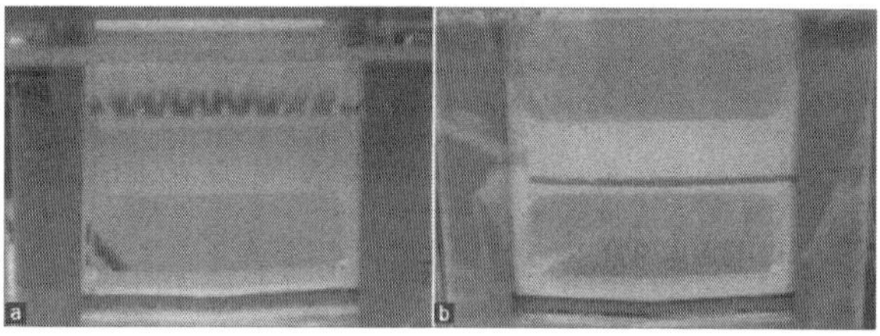

6. The gel is run until the dye front runs off the bottom of the gel as shown below in Figure 1.7 (Mahmood and Yang, 2012):

1.3.5 Electrotransfer

1. 6 filter sheets and one polyvinylidene fluoride (PDVF) membrane are cut to fit the measurement of the gel.
2. The sponge and filter paper are wetted in transfer buffer while the PDVF membrane is wetted in methanol.
3. The glass plates are separated and the gel is retrieved.
4. A transfer sandwich is created as follows:

Sponge

3 Filter Papers

Gel PVDF

3 Filter Papers

(Tip: make sure that there are no air bubbles between the gel and PVDF membrane).

Figure 1.8 below shows the sandwich (https://en.wikipedia.org/wiki/Electroblotting)

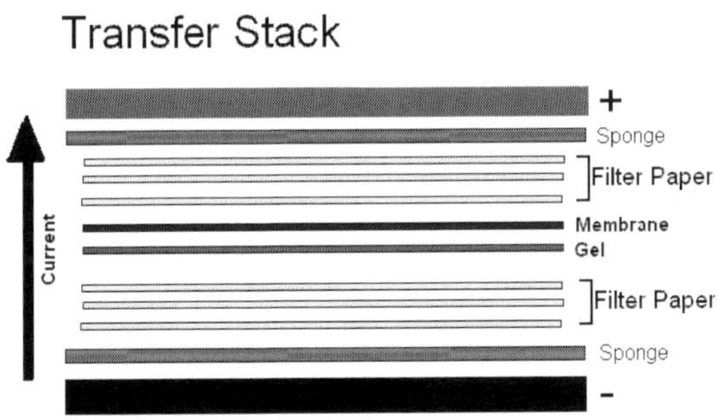

5. The sandwich is placed in the transfer apparatus, which in turn should be placed on ice to maintain 4°C. The transfer buffer is added to the apparatus and checked that the sandwich is covered with the buffer. The electrodes are placed on top of the sandwich, ensuring that the PVDF membrane is between the gel and a positive electrode. Figure 1.9 below shows the setup on ice (Mahmood and Yang, 2012):

6. Transfer is carried for 90 minutes (Tip: The running time is proportional to the thickness of the gel, so this time can become 45 minutes for 0.75 mm gels).

Figure 1.10 below shows the membrane after transfer (Mahmood and Yang, 2012):

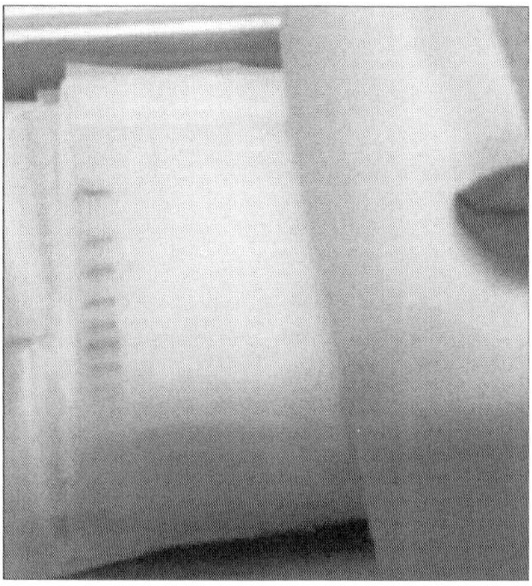

1.3.6 Blocking and antibody incubation

1. The membrane is blocked with 5% skim milk in TBST for 1 hour.

For TBST, the preparation involves:

Dissolve the following in 800 ml of distilled H2O

- 8.8 g of NaCl
- 0.2g of KCl
- 3g of Tris base

- Add 500ul of Tween-20
- Adjust the pH to 7.4
- Add distilled water to make the volume to 1L
- Sterilize by filtration or autoclaving

2. The primary antibody in 5% bovine serum albumin (BSA) is added and incubated overnight at 4°C on a shaker as shown below in Figure 1.11 (Mahmood and Yang, 2012):

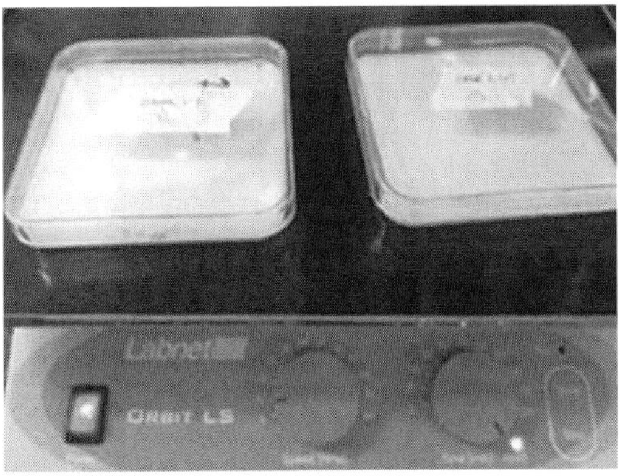

3. The membrane is washed with TBST for 5 minutes thrice. (Tip: All washing and antibody incubation steps preferably use shaker at room temperature to ensure uniformity).
4. Add secondary antibody in 5% skim milk in TBST, and incubate for 1 hour.
5. The membrane is washed with TBST for 5 minutes thrice.
6. The ECL (enhanced chemiluminescence) mix is prepared as instructed by the manufacturer. Incubate the membrane for 1–2 minutes. (Tip: Use a 1000 μL pipette to ensure that ECL covers the membrane completely).

Figure 1.12 below shows the addition of the ECL mix (Mahmood and Yang, 2012):

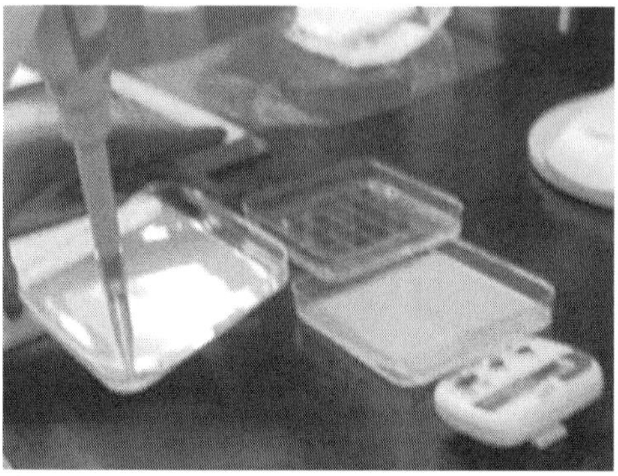

7. The results are visualized in the darkroom. (Tip: If the background is too strong, the exposure time is reduced). Figure 1.13 below shows the visualization in the darkroom (Mahmood and Yang, 2012):

(Mahmood and Yang, 2012)

As an alternative to commercially-available enhanced chemiluminescence kits, there have been reports of developing "in house" ECL detection kits to lower the cost. An example is Mruk and Cheng,2011. The developed ECL mix included:

- Luminol ($C_8H_7N_3O_2$, 5-amino-2,3-dihydro-1,4-phthalazinedione, also known as 3-aminophthalhydrazide); 250 mM luminol prepared in dimethyl sulfoxide (DMSO).
- *p*-Coumaric acid; 90 mM *p*-coumaric prepared in DMSO.
- H_2O_2 solution, 30% (w/v)
- DMSO
- ECL buffer; 0.1 M Tris pH 8.6 at 22°C

Figure 1.14 below shows the analysis of testis lysate using a commercial ECL reagent and the above-mentioned ECL mix (Mruk and Cheng,2011)

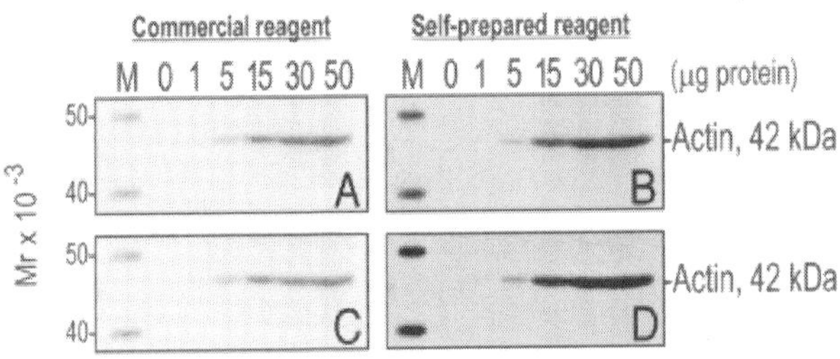

This self-prepared chemiluminescence reagent exhibited superiority to a commercially obtained kit in terms of sensitivity, signal duration, ease-of-use and shelf-life at a fraction of the cost of the commercial kit.

(Mruk and Cheng,2011)

1.4 TROUBLESHOOTING

A western blot experiment can encounter several challenges in terms of results: these include the following:

1.4.1 Unusual or unexpected bands

One cause can be degradation by proteases that cause bands to be formed at different than expected positions. This can be corrected by changing the antibody or using a fresh sample. In the case a protein has complicated structures or folding, reheating the sample can break the higher-order structures.

In the case of blurry bands, the culprits can be air bubbles that can be rectified by correct preparation of the transfer sandwich or changing the transfer buffer. Another cause can be a high voltage.

In the case of non-flat bands, rapid movement of samples through the gel can be the problem. This requires the correct optimization of the gel according to the sample.

1.4.2 No bands

This can be due to primary or secondary improper antibody. The concentration of the antibody must also be taken into account.

In the case the analyte is present at very low levels or not present at all then the use of another protein can give a picture of whether the problem is in the protein under study.

The signal can also be reduced due to excessive or long washing steps.

Another potential cause can be the buffers used: all the buffers used ranging from the transfer buffer, TBST, running buffer and ECL buffer should be fresh and free of contaminants.

In the case of a low concentration of antibody or antigen, the exposure time can be increased to make the band clearer. Alternatively, the blocking agent nonfat dry milk can mask the antigen/protein that can dictate lowering the quantity of milk used or use BSA.

1.4.3 High background

In order to lower the background in such Western blots the following points have been suggested:

- Tween-20 can interfere with the ECL reagent. This can be settled by rinsing the membrane briefly with Milli-Q or double-distilled water just before ECL detection.
- Before secondary antibody incubation, re-blocking the membrane with non-fat milk can lower the background. The determination of the concentration of milk and blocking time would depend on the experiment with 5–10% milk (w/v) for 5–10 min at room temperature is recommended.
- Washing with a high salt wash buffer [10 mM Tris pH 7.4 at 22 °C containing 0.5 M NaCl, 1 mM NaH_2PO_4, 0.2% SDS (w/v) and 0.1% Tween-20 (v/v)] for 30 min at room temperature with gentle agitation, followed by a brief rinse with Milli-Q or double-distilled water can be of help in the case of persistent high background.

1.4.4 Patchy and uneven spots

These are a result of the improper transfer. The presence of air bubbles trapped between the gel and the membrane can cause dark spots. This makes the use of a shaker for all the incubation steps to ensure uniform mixing.

Another cause can be the presence of contaminants in the blocking agents. This can be prevented by filtering the blocking agents.

In the case of aggregated secondary antibodies that can also cause this problem, the aggregates should be removed by centrifugation and filtration.

The overall flow and key points for successful Western blotting are presented in Figure 1.15 (http://onlinelibrary.wiley.com/doi/10.1111/sms.12702/full):

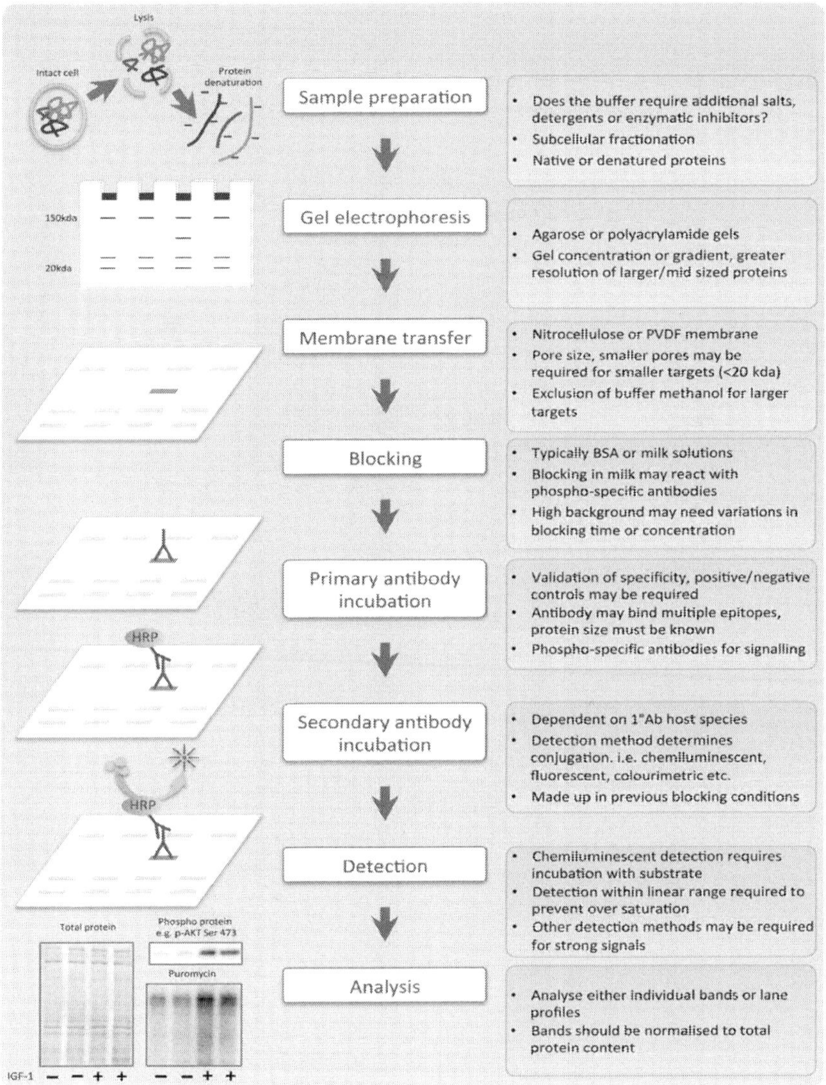

Table 1.2 below summarizes the various challenges in Western blots and their solutions (https://precisionbiosystems.com/western-blot-troubleshooting-guide/):

Issue	Possible cause	Solution
Diffuse Bands	1. Very high Antibody concentration 2. Excessive protein 3. Fast protein transfer was too fast and/or overheated gel	1. Lower the antibody concentration. 2. Lower the amount of total protein loaded on the gel. 3. Increase transfer time *and* Use a cooling system for electrophoresis.
Nonspecific bands	1. Non-specific binding to immobilized protein bands due to SDS 2. Protein expression profiles due to frequent passage of cell lines 3. Degradation of protein samples 4. Presence of new proteins or different splice variants 5. Presence of impurities in the antibody 6. Formation of multimers 7. Formation of different protein subtypes which have different molecular weights 8. Presence of a multi-modifier locus in protein 9. Very high primary antibody concentration 10. Non-specific bands caused by secondary antibody 11. Excess protein on the gel 12. Insufficient washing 13. Blocking problem	1. Wash blots after transfer and Don't use SDS 2. Use the original non-passaged cell line and run the original cell line samples in parallel. 3.Use fresh sample protease inhibitor. 4. Perform BLAST search: check the literature for other reports and use the cell line reported. 5. Use a monoclonal antibody. Purify the antibody by affinity method. 6. Before running SDS PAGE gel, boil protein for 10 minutes to disrupt multi-mer. 7. Study literature and use bioinformatics analysis to estimate the correct size of the protein. 8. Study literature and use bioinformatics analysis to confirm the presence of modifier locus and its molecular weight. 9. Decrease the primary antibody concentration 10. Use fluorescence labels on the primary antibody. 11. Lower the amount of total protein loaded on the gel. 12. Increase washes 13. Increase blocking time optimize the choice of blocking agent.

Weak or no signal from the blot	1. Missed the detection step or the detection reagents not working 2. Insufficient incubation with the detection reagent. 3. Poor or incomplete transfer 4. Protein of interest ran off the gel 5. Incorrect reagents added or incorrect containers are filled 6. The sample is very dilute 7. The protein is weakly bound to the membrane 8. Inactive or overly dilute primary or secondary antibody.	1. After the blot processing is complete, perform the detection step using your standard detection reagents and protocol manually. Make sure that the detection reagents are functional. 2. Remove the blot from the detection reagent when the signal-to-noise ratio is in range. 3. Make sure the transfer apparatus and membrane sandwiches are assembled optimally. Use appropriate transfer times. After blotting, stain membrane to measure transfer efficiency. 4. Use positive control and/or molecular weight marker . After blotting, stain the membrane to measure transfer efficiency. 5. Ensure primary and secondary antibodies, substrates, enzyme system and samples are compatible. 6. Load a higher concentration of protein onto the gel. 7. Use membranes with appropriate binding capacity. Dry PVDF membrane after protein transfer to ensure strong binding.
High background on the blot	1. Film overexposed or became wet during exposure 2. Short blocking time or washing intensity 3. High concentration of primary and/or secondary antibody 4. Protein is overloaded 5. Membrane, solutions, trays, or antibody containers are contaminated.	1. Decrease exposure time or allow the signal to further decay. Prevent leakage of solutions by encasing membrane in transparency film and blotting excess substrate from edges before exposure. 2. Increase blocking time and number of washes. 3. Determine the optimal antibody concentration by performing dilution. 4. Reduce the load or dilute the concentration of the sample. 5. Use clean glassware and purified water to prepare solutions. Wear clean gloves at all times. Use forceps when handling membranes. Clean the cleaning buffer two times.

High Non-specific binding	1. Insufficient removal of SDS or weakly bound proteins from the membrane after blotting. 2. Short blocking time 3. Affinity of the primary antibody for the protein standards 4.Overloading of protein	1. Follow the proper protocol. 2. Increase the blocking time 3. Check with the protein standard manufacturer whether the homology with primary antibody. 4. Reduce load or dilute the concentration of the sample

(Mruk and Cheng,2011; Mahmood and Yang, 2012)

1.5 ADVANCES IN WESTERN BLOT

1.5.1 Simple Western™

In order to augment the reproducibility, the timing and accuracy of western blot; the Simple Western is a reinvention of the entire Western blot. This involves the automation of steps ranging from the protein loading and resolution, probing, washing, detection as well as quantitative analysis. Once the samples are prepared as done for the regular western blot, the simple Western Sample Buffer and standards (final concentration: 1 μg/μL) are added followed by reduction and denaturation.

These prepared samples are mixed with primary and secondary antibodies and chemiluminescent substrates are added to low-volume 384-well assay plates. The other steps are carried out under automation: the separation of proteins happens in capillaries through a stacking and separation matrix. This is followed by immobilization of the separated protein is via photoactivated capture. This is followed by the identification using a primary antibody and subsequent immunodetection using horseradish peroxidase (HRP)-conjugated secondary antibody and chemiluminescent substrate.

This analysis of 12 samples can be done at a single time over 3–5 hours. For each protein, the molecular weight, area, percent area and signal to noise is presented. The use of automation allows for increased reproducibility as the various stages involve minimum human-induced variability. The reproducibility of the Simple Western assay compared to Western blot is shown below in Figure 1.16 for detection of phosphoinositide 3-kinase

(PI3K) expression in an MCF-7 lysate; a represents Simple western and b the Western blot (Nguyen et al, 2011):

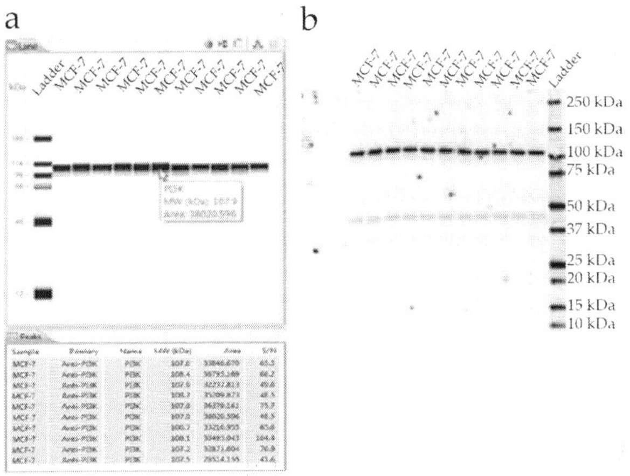

Figure 1.17 below shows the range for the Western assay glycogen synthase kinase- 3 (GSK-3) in a K562 lysate and Western blot for the same; the linear response was more in the case of the former. (a represents Simple Western and b represents the standard Western blot) (Nguyen et al, 2011)

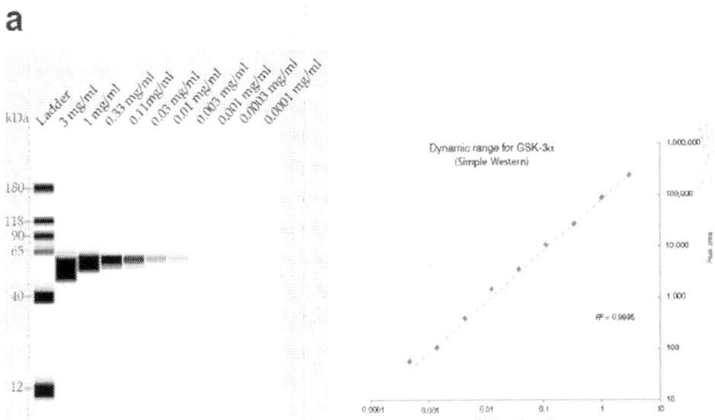

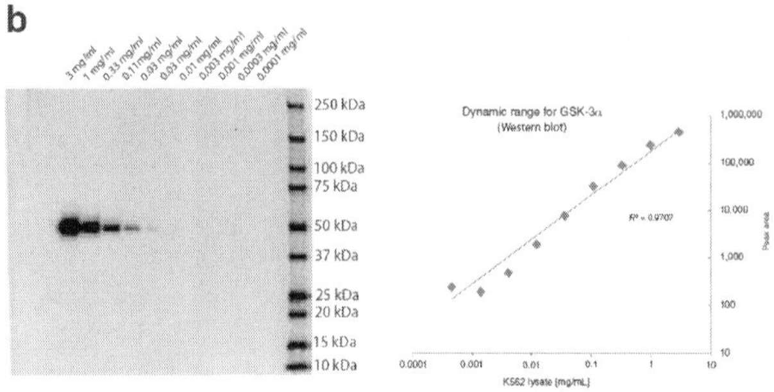

(Nguyen et al, 2011)

1.5.2 Multiplexed western blotting

A team published an article in 2010 using the regular procedures to resolve NIH-3T3 cell lysates using SDS-PAGE followed by transfer onto a PVDF membrane. Ponceau S was used to stain the proteins after which a microfluidic chip was sealed with the membrane. This chip had seven co-linear channels (150 μm wide × 100 μm deep and 3.5 cm in length) with inlet and outlet ports fabricated in PDMS. These channels overlayed the protein bands with each channel engineered to deliver a primary antibody probe solution permitting multiplexed labeling. The peeling of the PDMS chip was done followed by subsequent detection of the membrane using a labeled secondary antibody. Thus, 7 proteins could be detected simultaneously (Pan et al, 2010). This was further developed to boost the throughput with seven parallel channels in the PDMS chip. The probe antibody added to each microchannel can thus, interact with each lane making western blotting multiplexed and also use lesser amounts of antibodies than conventional western blots (He et al, 2015).

Figure 1.18 below shows the scheme of multiplexing: (A) Blotted PVDF membrane showing the electrophoretically separated proteins and ladder markers (B) Overlaying the protein bands is a microfluidic network of seven parallel serpentine channels against the PVDF membrane (C) Multiplexed detection with antibodies added to leach lane across each protein band (He et al, 2015; Sanders et al, 2016):

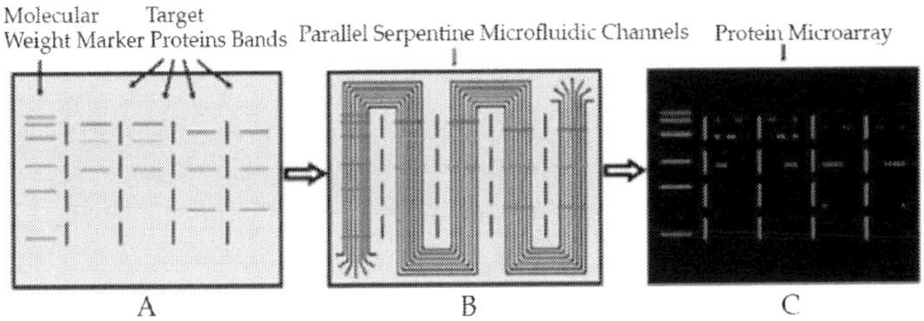

1.5.3 Changing the electrophoresis

The integration of capillary gel electrophoresis (CGE) with western blotting presents several advantages: CGE involves a polymer sieving matrix in small capillaries for separating the proteins treated with SDS. This not only lowers the sample volumes but also increases the efficiency of resolution.

For example, Kennedy et al (2011), combined SDS-CGE into a WB format-the polymer gel was filled in the separation capillary with a sheath around the outlet to prevent the formation of bubbles. This sheath was lightly connected to the PVDF membrane followed by voltage application. This led to the direct deposition of the separated protein bands from the gel on the PVDF membrane. This membrane with deposited protein can be used for conventional detection. This SDS-CGE direct blotting approach resulted in a lower time of analysis: less than 60 minutes for the protein lysozyme (MW=14 kDa); detection limit: 5 μg/ml (Kennedy et al, 2011).

Figure 1.19 below shows the SDS-CGE direct blotting system: Separation of proteins in a capillary filled with a sieving matrix using a negative high voltage at the inlet (A) and ground (D). Sheath flow was introduced at (B) followed by direct blotting of the separated bands (C) onto a PVDF membrane. (E) shows the separation capillary tip and the surrounding sheath capillary (Kennedy et al, 2011; Sanders et al, 2016)

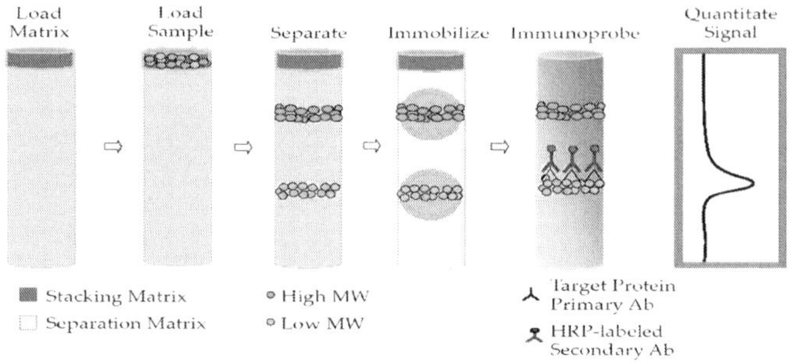

1.5.4 Overcoming transfer

In order to overcome the replica transfer step of proteins from a PA gel to a more porous membrane that does not permit the direct detection of antibody probes, the introduction of a protein immobilization step makes the process easier. Proteins separated using capillary electrophoresis were captured onto the walls of the capillary using photoactive moieties lined along the capillary. This system has been commercialized in SimpleWestern and ProteinSimple with proteins separated based on molecular weight were immobilized to the capillary wall using UV-based chemistry. Primary and secondary antibodies are then applied followed by chemiluminescence analysis from HRP linked secondary antibodies. Simple Westerns have shown to be more reproducible and lower the dependency on human operation thus increasing the throughput.

Figure 1.20 shows the scheme of the automated SimpleWestern method (Kurien and Scofield, 2015; Sanders et al, 2016)

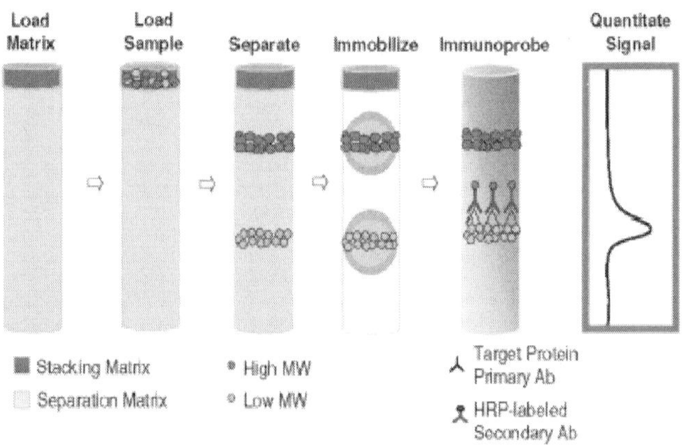

1.5.5 New detection systems

Other new developments in detection include Surface plasmon resonance (SPR) that allows quick measurements when coupled to capillary electrophoresis. Another option is the use of optical microresonators that have been integrated with separations and are label-free. An example is the use of whispering gallery mode (WGM) resonators that can trap and recirculate light to detect changes in the local refractive index due to the binding of an antigen and antibody.

An example is Kim et al (2016) who used capillary electrophoresis coupled to end-column WGM refractive index sensing. A Dove prism channelizes light from a diode laser into a barium titanate spherical microresonator in the lumen of the separation capillary. A photomultiplier tube (PMT) detected the light scattered by the microsphere as shown in Figure 1.21 below (Kim et al 2016; Sanders et al, 2016):

(A) the set up (B) an enlarged view

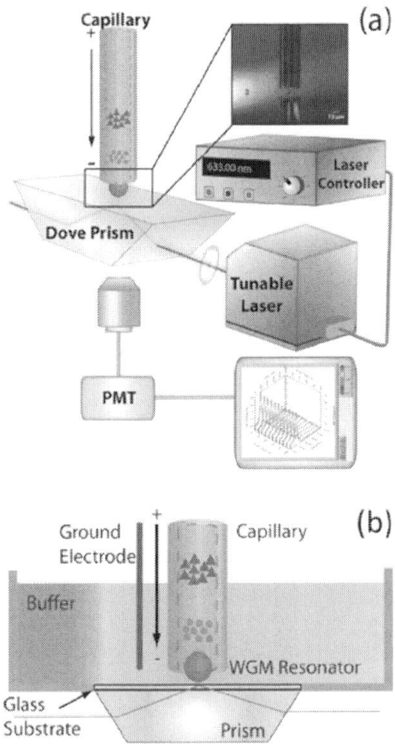

(Sanders et al, 2016)

1.6 FINAL WORD

The western blot had its origin in the 20th century; 1979 to be specific. It is now a ubiquitous member of a research lab. The technique has seen innovations in terms of the use of capillary electrophoretic systems and microfluidics instead of the traditionally used slab gels and electrotransfer. These systems have lowered the sample volume to the level of the single cells along with higher throughput. Such multiplexing of this traditional technique can allow new research to explore the nuances of our intricate biological systems.

1.7 REFERENCES

1. Brittany J. Sanders, Daniel C. Kim, and Robert C. Dunn. Recent Advances in Microscale Western Blotting. Anal Methods. 2016 October 21; 8(39): 7002–7013. doi:10.1039/C6AY01947A
2. Mahmood T, Yang PC. Western Blot: Technique, Theory, and Trouble Shooting. North Am J Med Sci. 2012; 4(9):429–434.
3. Mruk DD, Cheng CY. Enhanced chemiluminescence (ECL) for routine immunoblotting: An inexpensive alternative to commercially available kits. Spermatogenesis. 2011; 1(2):121–122.
4. Nguyen U, Squaglia N, Boge A, Fung PA. The Simple Western™: a gel-free, blot-free, hands-free Western blotting reinvention. Nat Methods. 2011.
5. Pan WY, Chen W, Jiang XY. Microfluidic Western Blot. Anal Chem. 2010; 82:3974–3976.
6. He S, Zhang Y, Wang P, Xu XZ, Zhu K, Pan WY, Liu WW, Cai KY, Sun JS, Zhang W, Jiang XY. Multiplexed microfluidic blotting of proteins and nucleic acids by parallel, serpentine microchannels. Lab Chip. 2015; 15:105–112.
7. Anderson GJ, Cipolla CM, Kennedy RT. Western Blotting Using Capillary Electrophoresis. Anal Chem. 2011; 83:1350–1355.
8. Kurien, BT, Scofield RH. Western Blotting: Methods and Protocols. Humana Press; New York: 2015. p. 465-467.
9. Kim DC, Dunn RC. Integrating Whispering Gallery Mode Refractive Index Sensing with Capillary Electrophoresis Separations Using Phase Sensitive Detection. Anal Chem. 2016; 88:1426–1433.

2

Southern Blotting and Associated Techniques

Shivsanjeevi Sripathi

ShivSanjeevi Sripathi completed his Masters in Biotechnology from Mumbai University in 2008. He was awarded for academic excellence in both his Bachelors and Masters for securing second rank in Mumbai University in 2006 and first rank in his college: Kishinchand chellaram College. For his Masters he secured first rank in his college KET's V.G.Vaze College. He qualified CSIR and NET and and TOEFL in September 2008. He then worked on a stem cell project at the Specialized Centre for Cell Based Therapy (SCCT),KEM Hospital at Mumbai on a project entitled, : Isolation & detection of stem cells from Human Umbilical cord/ amniotic membrane" following which he worked at Junior Research Fellow at Microbiology & Cell Biology Department, Indian Institute of Science, Bangalore from August 2009 to November 2011. The field of work involved cloning of cell wall genes and transcription factors in *E.coli & M.smegmatis*. As a writer, he has authored and co-authored 25 books with Delve and Arcler International Press. He loves to read and share on interesting aspects of life sciences in books. In his free time he loves to travel and explore ancient customs and traditions.

2.1 INTRODUCTION

In the field of biology, one of the predominant techniques is Southern Blotting. Southern Blotting owes its name to the first inventor, E.M.Southern, who, in 1975, devised a technique whereby DNA molecules, mostly restriction fragments, that are in an electrophoresis gel, get transferred to a nitrocellulose base nylon sheet, often called as the "membrane", without displacing the DNA banding that occurred in the electrophoresis gel.

A technique that is used to transfer DNA molecules to a Nylon membrane (nitrocellulose) from an electrophoresis gel, is what is termed as Southern Blotting, and is conducted before the identification of specific molecules vide hybridization probing.

Hybridization Analysis, where individual restriction fragments of DNA are targeted by specified DNA or RNA probes, the nylon membrane can be used as a substrate in case the source DNA becomes immobilized, either during transference or during treatment later. A restriction fragment detection in the background of several restriction fragments is the technique that Southern Blotting is all about (Brown, 1999). Southern Blotting is a precursor to RFLP Analysis – (Restriction Fragment Length Polymorphism), and Southern Blotting may very well be used to confirm the identity of a cloned fragment, or a sub-fragment within said cloned DNA, could be a bacteriophage clone or could be a plasmid even. This article will describe the standard principles of Southern Blotting, will overview the DNA preparation techniques, techniques for the actual blotting and the hybridization analysis as well (Brown, 2001).

2.2 INTERESTING INSIGHTS: THE ROLE OF DR. SOUTHERN IN SCIENCE

In 1973, Dr Edwin Southern developed a DNA blotting technique and ushered in standardized laboratory procedures in molecular biology that is in use today, several decades later. This has become the final word in the study of DNA structure. In the field of open science, the invention of DNA Blotting was distributed freely and widely among the scientific community and still stands as a base for study. The invention of the Southern Blotting Technique happened at the fortitudinous time when opinions on research on health and the immense value of molecular genetic study and research were becoming apparent in the industry. This meant that scientists went from basic research to working on the commercial applications of molecular biology.

Patented technologies such as the microarray methods, which came later, when compared with the unpatented Southern Blotting technique gives an indication of the evolution of the attitude of the industry, the people and the academicians towards commercialization of molecular biology which was largely academic leading to vast changes in the legislative policies of the United Kingdom to be more friendly towards commercialization of genomic research. DrSouthern's personal experiences in themselves are an illustration of the way things have changed perception wise towards commercialization and the adaptation of the scientists' attitudes about the same.

In 1962, Dr. Edwin Southern received his Ph.D. from the University of Glasgow, in the field of Biochemistry, and earlier studied Chemistry at Manchester University and progressed to study Nucleic acids. In 1967, he enrolled in the MRC Mammalian Genome Unit in Edinburgh after a four-year stint at the Low Temperature Research Station, the Antarctic Research Centre situated in Cambridge. This institution planted the seed of the pioneering new study that would eventually carry the term genomics, the Mammalian Genome Unit had several brilliant people who would later go on to become leaders in this new field. One of the many people whom Dr. Southern identifies as mentors, Professor Sir Kenneth Murray would go on to become a frontrunner in genomics research, is an example. Sir Murray co-founded Biogen, a biotechnology company that would go on to become a pan Atlantic entity. Sir Murray also co-discovered the Hepatitis B antigen.

The Mammalian Genome Unit fostered an open collegiate type environment with open sharing being encouraged, which led to accelerated inventions and discoveries. Dr. Southern believed that knowledge was not fostered by competition and therefore he saw to it that information passed readily between laboratories. A Club that was active during his tenure at MRC, the Restriction Enzyme Club, was a prime example of this approach. And to quote him:

'People were very keen to share. Just to give you one example, when restriction enzymes first came out… Ken Murray in Edinburgh set up a club to make restriction enzymes. To join the club, you had to make one, and you had to share the one you made with everyone else. It was very much a culture of sharing. And, I think as far as technology is concerned, whenever anybody had a technique in their lab, they were very keen to teach other people to apply it in their own work.'

To quote Dr. Southern, "Necessity is the Mother of Invention", and "Laziness is the Father of Invention", were two phrases that actually influenced his invention process.[7]In order to study the transcription of Ribosomal RNA from DNA, Dr. Southern needed to isolate the 5S Ribosomal RNA or rRNA genes, and this requirement led to the development of the new Blotting Technology. Density Gradient Centrifugation was the technique deployed at that time to purify genes and it was vastly inadequate. By separating the fragments based on size through gel electrophoresis, after isolating the restriction fragments by cutting the DNA strands with a restriction endonuclease, Dr. Southern managed to isolate these genes. After the process of separation was performed, the next challenge was the extraction of these separated genes from the gel. To quote Dr. Southern, "the prospect of cutting lots of gels into hundreds of slices, eluting the DNA from each of them and binding this to a filter that would then be hybridized and counted drove me to think that there had to be a better way of doing it."And it is the search for this new "better way" that the new Blotting technique was developed by Dr. Southern.

The utility of Southern Blotting as immediately obvious the moment it was explained. Separation of DNA fragments is a minimum prerequisite in every molecular biology procedure and the advent of Southern Blotting made this procedure extremely simple and quick. Among several structural analysis applications of DNA, Southern Blotting was used for clone genes study and study of gene expression.

The publishing of the invention by Dr. Southern was delayed. Though the invention was completed in 1973, his publication did not see print till 1975. The Journal of Molecular Biology rejected his paper, initially on the grounds that what was submitted was a paper that described methods and that the Journal would publish only procedures with interesting result showcases. Therefore, in order to demonstrate that his invention was indeed Journal worthy, Dr. Southern was forced to conduct several experiments.

Figure 2.1. Example of a pre-publication sketch for the Southern blot (Courtesy of Edwin Southern)

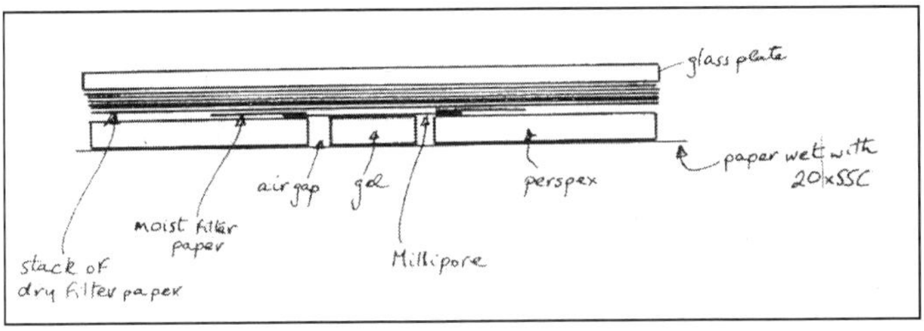

Dr. Southern started sharing much before his work got published. During his experiment running phase where he was gathering more data for the purpose of bolstering his publication, he engaged in quite a wide pre-publication strategy for sharing. The sharing process of the Sothern Blotting Technique started with Michael Mathews a Cold Spring Harbor Lab scientist, who took an immense liking to Dr. Southern's work after a visit to the Mammalian Genome Unit and expressed his interest in taking the technique back to the USA so that he can apply it on his own study of viruses. Dr. Southern gave away the schematic literally drawn on a piece of paper, which as per Mathews's description, went back to Cold Spring Harbor "clutching this drawing" (see Figure 1). Along with his colleague Mike Botchan, Michael Mathews implemented the Southern Blot and improved the technique, back in the USA. And that was when news about this hot new Blotting technique started to spread far and wide among the scientific community.

The demand for the procedure was obvious and the people at Cold Spring Harbor requested that they disseminate the technique further and farther. Dr. Southern agreed with the only precondition that he be acknowledged for this process origin. Dr. Southern still believes in the pre-publication strategy of his and says that "I believe that, if it had been published in the normal way, it would not have had such a firm attachment." When asked about the open sharing of his technique, he says, "I've obviously no regrets about doing that."

The "for attribution" option in the Creative Commons Licence bears a lot in common to Dr. Southern's open sharing of the technique with a hard insist on attribution. The motive of bringing similar openness in the publications with proper attribution has been the founding story of Creative Commons which was established in 2001. The creator of various disciplines is intricately helped by Creative Commons with regard to the publication of their works. The section "for attribution" is extremely popular among creators, as it permits free and open sharing with a guarantee that the acknowledgment of the original creator is always intact. The Southern Blotting Technique is a classic example of this attribution.

Under Margaret Thatcher in the 1980s, the management of Intellectual Property from publicly funded research projects changed form. "marked by the intensification of fears that British science was in serious trouble." was how one would describe the Thatcher years. The same strategy that Thatcher employed in economics, she brought on to bear in scientific research, meaning scale back the governmental role and let open market forces play a role in the scientific research arena. On 14 May 1985, the Thatcher government officially rescinded TC5/50 and abolished the BTG right of first refusal on patents. As a result, entities that used Public Funds for research like universities could now hold Intellectual Properties in their own name for in-house research. This change in the United kingdom's policy for granting private IP for research conducted with public funds paralleled the same type of changes across the pond in the USA specifically the Bayh-Dole statute of 1980.[26] The Bayh-Dole statute, in turn, reflected a growing U.S. belief that Public Funds diverted to research should be incentivized by Private IP and Licence Ownership with the belief that this would foster greater inventions and further value generations by the commercialization of the same. This change in outlook metamorphosed around the same time that commercial biotechnology was in its nascent stage, taking shape and fields of research such as cell fusion, recombinant DNA analysis and so on started gaining a lot of traction especially with regards to drug discovery, agriculture and diagnostics industries.

However, before getting the approval to apply for ownership of Intellectual Property, the research body should first show the UK Research Councils that it had the wherewithal for research and discovery and the ability to administer and manage its own IP. There were different approaches that the Public Research Councils took to give the applying Universities the right to apply for their own IP. After the 1999 Baker Report, there was much clarity that was born in the overall process. In the Baker Report, John Baker submitted recommendations to the Minister for Science and the Financial Secretary to the Treasury for "increasing the rate at which their [Public Sector Research Establishments] research is successfully commercialized." The government's response was that Research Councils should mandatorily disclose the specific disposition of intellectual properties of universities. This brought under control the distinctions and variance among various councils.

Oxford, where Dr. Southern now worked, jumped at this new opportunity to own patents and license applications in its own name, thanks to the Thatcher directives. Isis Innovation was formed in 1988 to manage and exploit university Intellectual Property as a direct result of the Thatcher Administration changes. IPs created at Oxford were held by Isis Innovation which actively transferred these technologies to the private sector for licensing fees and agreements. Professor Southern Was on the founding Board of Isis and therefore, had an intricate role in the commercialization process at Oxford.

Along with the marked changes in the United Kingdom's policies Professor Southern's policies regarding the sharing of technology and inventions also took a noticeable shift. The origin of the Southern Blot was at a Public Sector funded Research establishment, the MRC Mammalian Genome Unit at Edinburgh in 1973 when the government took the rights for inventions made with public funds. If the Southern Blot were to be patented, NRDC would hold the right to the patent, and hence it was not patented at all. The Southern blot was created before the 1985 Thatcher changes and was not patented; the microarray was created after the 1985 policy changes and was patented. Therefore, the policy changes seem to be the only deterministic factor in whether the patent was taken or not, and served as the principal decision.

In 1985, when Dr. Southern filed his first patent application for microarrays and in 1987 when he started work on Microarrays, the policy changes post- Thatcher was so dramatic that the IP Rights landscape had totally changed. Margaret Thatcher rescinded TC 5/50 in 1985, thereby allowing universities to not just own but also exploit Intellectual Properties, subsequently, several other laws had come into effect in the United Kingdom to foster commercialization and encourage private enterprise. A direct fallout of this was the Oxford University establishing a TTO called the Isis Innovation, aimed at exploiting the Intellectual Properties that arise as a result of innovations happening at Oxford. There is now a greater financial incentive to the patent at Oxford with a separate department being created just for the management of Intellectual Property Rights and motivates employees to report patentable inventions.

There is a marked change in the way Dr. Edwin Southern feels about science norms and Intellectual properties today when compared to the way he felt in the seventies. What has brought about this change and what are the factors responsible for this evolution? Let us look at Dr. Southern's accounts represented as that of the scientific community at large, as through the eighties the scenario prevailed as that of forceful acceptance of the Intellectual Property domain:

'I think gradually I became under pressure. We all had to, we were under contract, to tell our funding agencies of any inventions which were patentable, anything in our work that could be commercialized. And at first, we were very resistant to that attitude. We thought that this would do the things you have been hinting at, you know, inhibit the spread of ideas and materials and so on. And so there was a resistance to that initially. When did that change? I think by mid-80's it was beginning to change. People were beginning to accept the idea that we had to do it, and that eventually became a sort of matter of course. I don't know anybody now who would put up a really strong resistance to that idea. They may not like it, but they would conform.'

The above statement clearly shows the shift in Dr. Southern's attitudes towards intellectual properties from the seventies through the eighties.

"We are pleased that the English courts have agreed to decide this issue as a preliminary issue, and believe that they will declare invalid Affymetrix's attempts to secure a license by means other than negotiation with OGT." when we read through this OGT (Oxford Gebe technology) Press release we see a Dr. Southern who is more a businessman than a scientist. Dr. Southern acquired several roles as he became more invested in OGT and got to grow personally and financially in his foray not science and Intellectual property rights. Dr. Southern of the seventies was a scientist, a colleague, a friend and a laboratory manager. Today, he all these things plus a lawyer and a businessman rolled in one. The fundamental catalyst for such a sea change in Dr. Southern is the fact that he is an industry-savvy scientist against a core academic scientist (Tofano et al, 2006).

2.3 TECHNOLOGY TYPES

In terms of marketing potential and complexity, the Southern Blot and the microarray are notably different despite the evolving views of Dr. Southern with regard to patenting since the seventies. One principal reason why the Southern Blot is so widespread in its usage across all labs all over the world is the fact that it is easily reproducible with materials on hand. " A paper towel and some salt solution" is literally all that a Southern Blot requires. Contrarily the Southern Blot's simplicity also makes it impossible to be packaged as a profit maker. A pre-packaged profit included Southern Blot Kit is a hard sell especially because it is easily made with cheap components and not much effort.

Contrastingly, Microarray technology is overly complex and requires extensive instrumentation and complex hardware for detection and interpretation. Also, integrating the components for making microarrays requires complex multi-disciplinary engineering prowess and the final assemblies require substantial investments, and therefore, universities can rarely venture beyond the prototype stage. Economies of scale also come into play as well as the requirement for quality control, which makes this the ideal target for private firms to manufacture and distribute them effectively. The microarrays require thousands of small clean chips that have to be spotted with thousands more oligonucleotide strands that originally have to be synthesized, in contrast to the simple materials required for a Southern Blot. In order to make a microarray, there needs to be expensive equipment in the lab, populated by expert staff, with several microarray chips being consumed in experimentation. Therefore, small laboratories cannot produce microarray chips and the dedicated microarray companies are a total necessity for the research bodies that utilize microarrays. Southern believes:

'In the case of microarrays, to do it properly you have to commercialize it. I think the "home-brew" business of microarrays sort of works alright, but it's clear by now, I think, that the commercial flavors of microarrays are superior to the home-brew flavors. You might find people that would argue with that, but I think… I'm sure that's really true.'

Owing to the difficulty and complexity in the manufacture of microarray chips, it stands a greater potential in the market than the easy to manufacture Southern Blot. This leads to the fact that a patent on microarray technology would be far more profitable than any patent on a Southern Blot technique. Another important factor that weighs in on the profitability quotient of the patent is the fact that it can be enforced in the course of legal play. Since Southern Blots are easily replicated litigation and enforcement are exceedingly difficult. However, enforcement of legal actions with regard to patents in microarrays is easier and more realistic because a substantial laboratory is required to manufacture these and is not possible by small outfits. It is also better for manufacturers to license the patents rather than individual laboratories and simple lab methods. In effect, targeting larger entities in the case of litigation for enforcement of patent rights yields greater benefit than small and ineffective litigations.

A level of maturity and talent is required in balancing decisions on balancing business decisions and pure science while deciding on the course of technology. Therefore, expert analysis of the cost benefits of patenting as compared to the more liberal open sharing methods needs to be done. Dr. Southern reasons as follows with regarding the fact that despite applying for many patents, very few pay off:

'It's a startling statistic, the costs of filing a patent are quite high as you undoubtedly know. You've got to pay your lawyers, in Europe you've got to pay - to maintain the thing - you've got to pay translation costs, so the total cost of taking a patent through to the point where it is really useful is about £20,000 to £50,000. You can almost double that for dollars… That is serious money. To earn that sort of money, let's suppose it's something that could bring in a 10 percent royalty… To justify a 100,000-dollar investment in the filing of the patent, it's got to be a million-dollar invention. Just to break even. To make anybody any money, it's got to be a 10-million-dollar invention.

There aren't many of those around.'

The difference between the profitability with regards to the Intellectual Property of Microarrays as compared with the Southern blot is testimony to the evolving landscape with regard to biotechnology patents in recent years. According to Walsh and Cohen, "the revolution in molecular biology, combinatorial chemistry, bioinformatics and related fields… has spawned discoveries of enormous commercial value since the 1970s."Greater commercial value, ever-increasing complexities in manufacturing technologies and greater investments all over, contribute to the ever-increasing patents of molecular biological inventions leading to the changes in the scientific norms (Tofano et al, 2006).

2.4 THE SOUTHERN BLOTTING PRINCIPLE

Hybridization studies in the 1950s and the 1960s on various forms of nucleic acid had utilized bound DNA from nitrocellulose powder and sheets, meaning these abilities have been in use for many years. In the formative years, there was no fragmentation of the immobilized DNA, meaning un-fractioned DNA that was spotted on nitrocellulose sheets or bound to nitrocellulose powder, was used. The early 1970s saw the ushering in of electrophoresis gel as a medium, which translated into the ability of DNA restriction fragments, separated on the basis of their size, to be stored, led to the research and development of several techniques for transferring these separated fragments on to nitrocellulose support. E.M.Southern laid out a technique, in 1975, wherein DNA was transferred from an electrophoresis gel medium to a nitrocellulose sheet placed on top of it, using capillary action, and this was a highly effective yet extremely simple system, and there has been little embellishment to these techniques over the years.

Figure 2.2 illustrates the original technique of Southern Blotting. Fractionated DNA fragments that exist in an agarose electrophoresis gel, are kept on a filter paper wick, and this forms the conduit between electrophoresis gel and the vat of high salt buffer.

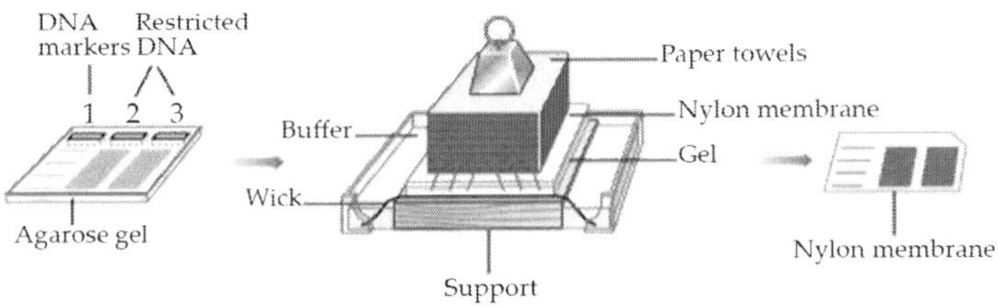

Figure 2.1. Southern blotting (Brown, 2001)

The nitrocellulose sheet is placed on top of the said gel and paper towels stacked in a tower form the technique and a weight is placed to hold this assembly in place. The buffer soaks through the paper wick, the electrophoresis gel, the membrane and into the paper towels through capillary action. The buffer carries the DNA fragments as it passes through the membrane, where they get attached to the nitrocellulose. Overnight, meaning approximately 18 hours, the transfer of fragments measuring up to 15kb length takes place effectively. There is a complication possible, wherein the buffer soaks straight into the paper towels from the wick, bypassing the gel, and this can be avoided if the process is set up with a lot of care (Brown, 2001).

2.5 RECENT ALTERATIONS TO SOUTHERN BLOTTING

While many improvements have happened over the years in the increase of efficiency of transfer of DNA from gel to a membrane, the technology is shown in Figure 1 is the exact representation of Southern Blotting as it is carried out in several laboratories. The advent of Nylon membranes has been a distinct advantage over nitrocellulose counterparts. Nitrocellulose sheets are extremely fragile and do not take to rough handling especially during the Southern Blotting process, also disintegrating when an attempt to perform more than one hybridization analysis using the same blot. Nylon membranes, on the other hand, are not fragile at all and can take all sorts of handling during the blotting process, also a single blot can be used for up to ten repeat hybridizations, the limit of ten being enforced, not because of the membrane's structural limit, but because of loss of blotted DNA over repeat hybridizations. Another advantage of nylon membranes is covalent bonding. Under certain conditions, such as a positively charged nylon membrane used with an alkaline transfer buffer, the transferred DNA gets covalently bound to the membrane.

Figure 2.3 below shows the scheme of Southern blotting (https://laboratoryinfo.com/southern-blot/):

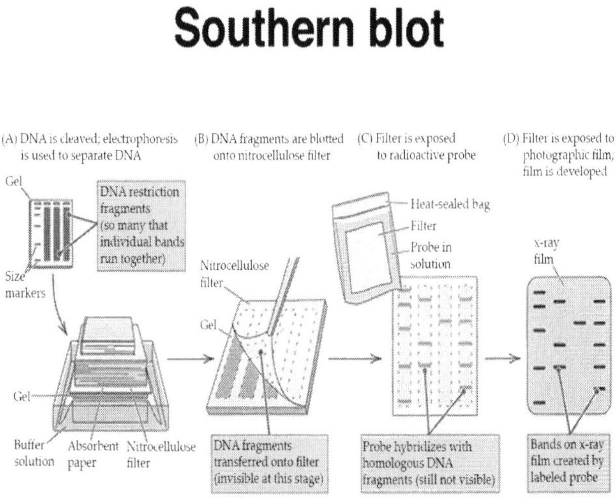

However, in the case of a nitrocellulose membrane, the DNA is bonded in a semi-permanent manner and immobilization occurs when the membrane is baked at 80 degrees Centigrade. Therefore, transfers happening to a positively charged membrane restrict DNA loss during the Blotting procedure as a result of leaking through the membrane; the process is also far quicker with the actual transfer duration being just 2 hours as compared with 18 hours. While nitrocellulose membranes effectively handle only molecules that are longer than 500bp, Nylon membranes can bind fragments as low as 50bp in length. There is a much lesser level of background hybridization, mainly when we consider probes that have been indexed with nonradioactive markers, in the case of nitrocellulose, and this is a reason why nitrocellulose is still in play when compared to nylon membranes.

There are more changes being researched and introduced incrementally to Southern Blotting over the years, in order to increase the efficiency of transfer. There have been suggestions towards alterations in the architecture of the capillary blotting system, which happen to be more complex than the simple structural arrangement of transfer upward, as laid out in Figure 1, and on top of it, none of the suggestions come with proportionally better improvement in either efficiency or speed. Non-capillary techniques, such as electroblotting, for example, show a lot of promise with regard to improvements, as they utilize electrophoresis to transfer the DNA from the gel to membrane. Originally developed for blotting polyacrylamide gels and also for agarose gels (Bittner et al., 1980), this technique was resorted to as the pore sized were too tiny for successful capillary movement of DNA. Since there is a palpable risk of bubbles being formed if the gel gets too hot at the time of transfer, this process of electroblotting has to be performed with adequate oversight. As far as agarose gels are concerned, vacuum blotting is a more sought to approach, as it uses the pressure or vacuum to move the buffer from gel to membrane, and the transfer process is also much quicker at a mere thirty minutes. Another variation to the technique of DNA extraction is to do away with Southern Blotting in entirety and to conduct hybridization analysis directly on the electrophoresis gel, by drying the gel and achieving DNA immobilization, by ensuring that the DNA is trapped in the gel matrix that is a dehydrated or solid support to which the DNA is bound before the gel is placed for drying. While this technique has proven useful for some, it does suffer from increased background signals subsequent to hybridization analysis (Brown, 2001).

2.6 DNA PRE-PREPARATION BEFORE SOUTHERN BLOTTING

2.6.1 Genomic DNA preparation

The type of DNA that is being researched determines the technique to be adopted for the preparation of said DNA for Southern Blotting. The goal, as far as Genomic DNA is concerned, is to isolate specific restriction fragments greater than 20kb in length from the molecules that have not been severely fragmented during random shearing during extraction procedures. It means that cells should be gently broken open, relatively speaking. High molecular weight DNA can be released easily, in the case of tissue and blood cultures and samples, by incubating them in a detergent buffer such as SDS – Sodium Dodecyl Sulphate, which is enough to disarray their cellular membranes. Bacteria which are usually encompassed in a cell wall can be gently opened up with a mixture of lysozyme and ethylenediaminetetraacetate (EDTA). Polymeric compounds in the bacterial cell walls are degraded by the enzyme Lysozyme and magnesium ions, which are crucial for the stability of the polymeric structure, are chelated by EDTA. Cells burst upon the addition of detergent. While plant cells are usually broken by physical grinding, after freezing them in liquid Nitrogen, cell wall degradation enzymes are used as deemed fit on microorganisms with tougher cell walls like fungi and yeast.

While some DNA gets fragmented, the greater yield is obtained. By taking measures to eliminate the main biochemicals other than the DNA, in the original extract, Genomic DNA extraction procedures can continue after cell disruption. Adding phenol or sometimes an even ratio of phenol and chloroform is usually used to conduct deproteinization, but a protease like Proteinase K can also be added to the buffer to jump-start the degradation of proteins in the extract. Subsequent to the centrifugation procedure, while the nucleic acids remain in an aqueous state, the proteins that have precipitated move to the interface in the middle of the organic and the aqueous state. Cetyltrimethylammonium bromide (CTAB) is commonly used as a secondary purification detergent, in the case of plant extracts, which have a greater volume of carbohydrates when compared to microbial and animal cells. During phenol extraction, CTAB particularly binds to the nucleic acids thereby improving the recovery efficiency. Low molecular weight contaminants like ribonucleotides and others are removed by precipitating the balance of nucleic acid and polymers, since several procedures for extracting include RNA digestion using ribonuclease and an end treatment with ethanol. After drying the DNA that is precipitated is then re-dissolved in a Tris-EDTA buffer or water.

2.6.2 Plasmid/phage

In order to get bacteriophage DNA or plasmid DNA from bacteriophage clones, unique techniques need to be used. Since plasmids are small supercoiled molecules when compared to genomic DNA which are long and linear fragments, post cell disruption, these physical differences can be targeted by several techniques in order to obtain plasmid DNA. One technique is where, by treating cell extracts with alkali, the linear DNA gets precipitated while supercoiled plasmids remain in the solution. Bacteriophage DNA, for example, recombinant l or M13 vectors, is usually got from bacteriophage particles directly and these particles are secreted by infected bacteria, and since these particles consist entirely of DNA and protein, the purification process gets very simple (Brown, 2001).

2.7 TECHNIQUES FOR DNA BLOTTING

After treatment with one or more restriction endonucleases, the purified DNA is then fractionated by electrophoresis by using agarose gel following which pre-treatment happens in the process of readying for Southern Blotting. The 2 goals of pre-treatment are as follows. Ideally DNA molecules, if they are broken into individual bands in small fragments inside the gel, the transfer will happen faster, as smaller fragments move through faster than larger ones. The way to achieve this is by soaking the gel in 0.25molL HCl for at least thirty minutes, yielding a minute amount of depurination – separation of the β-N-glycosidic bond among the two purine bases (adenine or guanine) and the sugar component of their nucleotides – following which the sugar structure starts decomposing and the polynucleotide chain starts breaking. The double-stranded DNA molecules getting denatured by an alkaline solution thereby breaking their hydrogen bonds, rendering the molecules single-stranded. This forms the second pre-treatment.

This helps the transfer and the bond with the membrane and post binding also ensures that the polynucleotides base-pairing components are available for probing for hybridization. Since DNA does not bind to nitrocellulose with a pH greater than 9.0, the pre-treatment must consist of the gel being soaked in a Tis-salt buffer, in case a nitrocellulose membrane is being deployed for the alkali pre-treatment. As shown in Figure One, by using a high-salt transfer buffer, the Southern Blot is set up, the buffer usually being the formulation called '20 SSC', which comprises 3.0molL21 NaCl (salt) and 0.3molL sodium citrate. Meanwhile, for transfer to a nylon membrane, the same alkaline transfer buffer(0.4molLNaOH) can be employed, but with a positively charged nylon membrane because, as explained earlier, the DNA transferred to the membrane gets attached immediately because of the covalent bond. In this case, the pre-treatment with alkali is not required. In the case of an alkaline blot, the blot sits for two hours and in the case of high-salt transfer, the Blot sits for at least eighteen hours. Subsequent to blotting, the membrane is rinsed in 2SSC after being dismantled and is left to dry. At this stage, the DNA is loosely bound to the membrane if the blot has been devised as nitrocellulose or as a nylon membrane that is uncharged. For more permanent immobilization to occur, baking should be done at eighty degrees centigrade for two hours thereby resulting in a semi-permanent DNA attachment to the nitrocellulose membrane, or covalent attachment of DNA to a nylon membrane by UV radiation.

The fundamental principle that when nucleotide sequences of two polynucleotides are totally or partially complementary, then a stable hybrid base-pairing, is the basis of Hybridization Analysis. Therefore, by probing the membrane with a labeled DNA molecule comprising of similar or the same sequence as the fragment that is desired, the required specific restriction fragment in the Southern Blot can be detected. Initially let us analyze the methods used for hybridization analysis and we can explore the different types of hybridization later.

2.8 THE SOUTHERN BLOT PREHYBRIDIZATION

The way Hybridization Analysis is done is either by soaking the Southern Blot in a buffer that is in a sealed plastic bag placed on a shaker, or alternatively a tube that is rotated so that all parts come in contact with the probe. There are two stages to Hybridization. 1) A solution that is designed to block the unused DNA binding sites on the surface of the membrane is used to pre-hybridize the membrane. The omission of this step will result in the probe binding non-specifically to the surface of the membrane and therefore identification of the signal for the specific restriction fragment will be difficult to identify. The prehybridization solution hence is made up of bovine serum albumin or dried milk, nonbiological polymeric compounds like polyvinylpyrrolidone and/or biological polymers such as Ficoll (a carbohydrate-based compound). An unrelated organism's DNA can also be used, when compared to the organism for whom the DNA is blotted (salmon sperm DNA is often used). Depending on the type of the membrane, prehybridization takes anywhere between fifteen minutes and three hours at 68 degrees centigrade.

2.9 WASHING STEPS AND HYBRIDIZATION

The hybridization process itself is the second stage, and this is done in a high salt buffer with a detergent -like 11% SDS. In this phase of the procedure, there are two very vital issues. Firstly, to be commensurate with the label carried by the probe, in order to be observed by the system, sufficient probe DNA must hybridize to the restriction fragment that is the target, to generate a clear signal.Mostly even if the probe has been labeled to the maximum for the specific activity(4109 dpmmg21 fora 32Pradioactivelabel), and the maximum quantity of genomic DNA is extracted onto the electrophoresis gel (usually, about 10mg of DNA per lane)there may be issues in achieving sufficient sensitivity for demanding applications such as the detection of a single-copy gene from human genomic DNA. In case a radioactive label such as 35S is used, and this is used only mostly in very non-complex DNA strands such as bacteriophage clones or plasmid clones, while ideally the radioactive label that should optimally be used being 32P, or even if a non-radioactive probe is used, more number of problems are likely to arise.

In such cases, an addition of an inert polymer like 10% dextran sulfate or 8% polyethyleneglycol6000 in the hybridization solution, can result in some increase in the sensitivity. The belief is that probe molecules are being induced by these polymers to form networks, thereby enabling larger volumes of DNA to hybridize to the membrane's target sites. In an actual use case, a ten to hundredfold increase in sensitivity can be achieved (Amasino,1986). The specificity of the reaction is the second critical step in the factors that should be considered for the hybridization.The importance of choosing the probe DNA section carefully means that there will be a portion in the selection that will be similar to the blotted restriction fragment, which is what is required. In case the region of hybridization is not completely congruent with the target, then it should at the very least possess strong similarities with the target region so that bonding can occur. There is an issue in that the probe itself has the propensity to bond with other blotted DNA fragments with which it has partial similarity.

Even if the best match is with the target that is specified, there can be some non-specific target problems with regard to hybridization, because even if the match is more complementary with the second restriction fragment than it is with the principally sought restriction fragment, the probe will not return an unambiguous signal. What this means is that a "Stringency" step should be carried out during hybridization to ensure that the probe to target hybrid is the one that remains stable while the other hybrids are unstable, and this stringency is achieved by evaluating exactly the composition of the buffer and the temperature of the hybridization process. The composition of the buffer is vital for maintaining the stability of the hybrid because the relevant factors for maintaining stable hybrid bonds are the ionic strength of the bonds ad the extent of occurrence of the destabilizing agents such as the hydrogen bonds disrupting formamide. In the case of a fully paired hybrid the melting temperature, Tm, meaning the highest temperature at which the hybrid bond is still stable, is always higher than the hybrid with base pairs that have been created with not fully complementary Target to DNA sequences.Thus, by maintaining proper hybridization temperature and the ideal composition of the buffer, destabilizing non-specific bonds can occur while the formation and maintenance of the required hybrid are achieved.

There are many strategies to achieve this goal. For instance, in the case of a cloned restriction fragment, assuming the length of the probe is 4100bp, by deploying the stringent conditions wherein the baking is at 68 degrees centigrade in a high-salt buffer, there is no way non-specific hybridization an occur, while the desired stable hybrid bonds are almost guaranteed. When the procedure is completed under these stringent conditions the steps for washing, after the completion of the procedure, will be merely to remove the probe that is non-hybridized. But then, in case fifteen to twenty five nucleotides in length, the short oligonucleotide probe is deployed, then by conducting the hybridization in a low stringency environment, wherein lesser temperatures than the prescribed Tm for the target hybrid, are maintained to allow for the formation of all hybrids along with the nonspecific ones. By conducting a series of washes in increasing temperatures after the hybridization process, hopefully, the desired specificity is what remains at the end of the procedure. A detection experiment that is relevant for the label, for instance running autoradiography for a radioactive label, is performed on the membrane post-washing. In case there is a requirement for subsequently re-probing the membrane, it can be stripped of DNA by washing it in a high -temperature alkali and detergent buffer which will effectively destabilize the hybridized DNA. The fact that nylon membranes with covalently bound

DNA strands can be washed only around ten times, and the fact that some blotted DNA are near impossible to wash off, renders this procedure less than optimal (Brown, 2001).

2.10 ROLE OF SOUTHERN BLOTTING IN IDENTIFICATION OF CLONES

The projects that aim at the cloning of a specific gene, and hence its identification, form the most common application of Southern Blotting procedures. With regards to the specific gene that is being sought, Southern Blotting aims at the identification of one or more restriction fragments in the genomic DNA that contain the gene, and subsequent to cloning and identification of the specific recombinant gene by either plaque hybridization probing or by the colony, the restricted cloned DNA is then used to identify, hopefully, a shorter restriction fragment, containing the target DNA sequence, from within the cloned DNA itself. Southern Blotting is vital because, while the target genome sequence may just be a few kb in length it may be ensconced in several kb lengths artificial chromosomes of some high capacity vector like a bacteria or yeast. Therefore, in order to be effective for genetic applications, Southern Blotting will have to have the most suited hybridization probe, which would detect one or more specified restriction fragments.

Sometimes, an existing cloned DNA can be deployed to ascertain the genomic version for that specific gene. In the case where the target gene is actually cloned from an organism similar to the target organism, then the usage of heterologous probing is possible wherein non-complementarity between the target and the probe can be tolerated. This technique can be used to find homologs of human genes in the genetic structure of other primates and can even cross the species barrier, for instance, the usage of Drosophila genes as probes on related human structures. Another technique involves the usage of amino acid sequences for designing a mixed nucleotide probe with the code of the target gene, that would yield in a variety of similar sequences that when put together would code a short sequence of the amino acid. For instance, the mixed oligonucleotide 5'-AA(C/U)GA(A/ G)AUGUU(C/U) UGGUA(C/U)GG-3', which contains 16 different sequences, covers the entire spectrum of possibilities for the amino acid sequence asparagine-glutamine-methioninephenylalanine-tryptophan-tyrosine-glycine and will most likely detect the DNA sequence coding when data is used adequately in a high stringency Southern Blotting procedure.

2.11 POLYMORPHISM ANALYSIS OF RESTRICTION FRAGMENT LENGTH

RFLP Mapping, or Restriction Fragment Length Polymorphism technique of Southern Blotting is a vital application in hybridizations in constructions of genome maps (Brown, 1999). The same restriction enzyme from different individuals does not always generate the same set of restriction fragments mainly due to the fact that most restriction sites are present in some and not in others, meaning they are polymorphic, usually just a point change in a nucleotide sequence will render it unrecognizable by the restriction enzyme. The construction of the first comprehensive human genome maps owed their creation to these polymorphic sites as they were used as markers in genetic morphing (Donis-Keller et al., 1987), there are about 105 RFLPs in the human genome. The different types of Restriction Fragment Length Polymorphisms as displayed by Southern Blotting are shown in Figure 2, a DNA fragment probe that spans the polymorphic region (possibly one of the polymorphic fragments, possibly a fragment prepared by cutting the DNA with a different restriction enzyme, or possibly a fragment obtained by PCR) and the presence or absence of the polymorphic site determined from the size(s) of the restriction fragment(s) detected after Southern blotting.

Figure 2.3 Using Southern Blotting to detect a Restriction Fragment Length Polymorphism (https://www.ncbi.nlm.nih.gov/probe/docs/techrflp/)

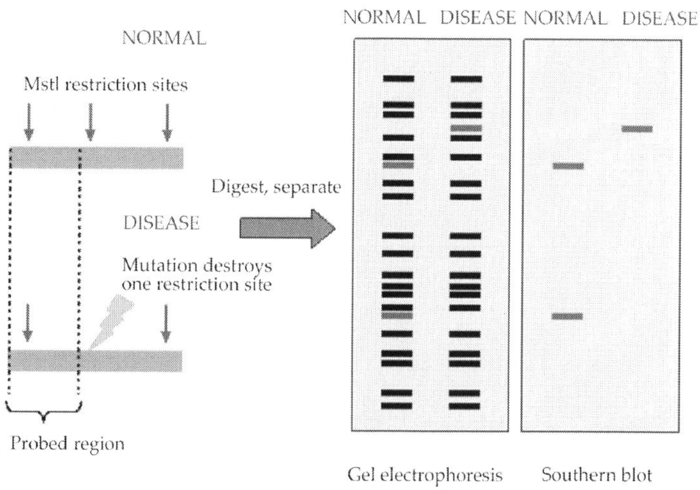

(Brown, 2001)

2.12 FINAL WORDS

Among a plethora of restriction fragments, the detection of a specific restriction fragment is what constitutes Southern Blotting. The way that Southern Blotting works is in the transfer of DNA fragments from an electrophoresis gel to either a nylon membrane or a nitrocellulose medium, wherein the pattern of DNA fragments that are present in the gel get replicated exactly in the membrane. The specific restriction fragment that is being sought after is then obtained y hybridization probing. While modifications, in order to achieve greater efficiency and faster transfer speeds, have come to force, the basic principle of Southern Blotting remains the same since its invention in 1975. The detection of Restriction Fragment Length Polymorphism, which is invaluable in the creation of genome maps, and the identification of the restriction fragments and DNA sequences of research interest have been some vital parts of molecular biology that have benefited tremendously because of the application of Southern Blotting.

2.13 REFERENCES

1. Terence A Brown (2001), Southern Blotting and Related DNA Detection Techniques, Encyclopedia of Life Sciences, Manchester, UK.
2. Tofano, D., Wiechers, I. R., & Cook-Deegan, R. (2006). Edwin Southern, DNA blotting, and microarray technology: A case study of the shifting role of patents in academic molecular biology. *Genomics, society, and policy*, *2*(2), 50. https://doi.org/10.1186/1746-5354-2-2-50
3. Amasino RM (1986) Acceleration of nucleic acid hybridization rate by polyethylene glycol. Analytical Biochemistry 152: 304–307.
4. Bittner M, KupfererPandMorris CF (1980)Electrophoretic transfer of proteins and nucleic acids from slab gels to diazobenzyloxymethyl cellulose or nitrocellulose sheets. Analytical Biochemistry 102: 459–471.
5. Brown TA (1999) Genomes. Oxford: BIOS Scientific Publishers. Donis-Keller H, Green P, Helms C et al. (1987) A genetic map of the human genome. Cell 51: 319–337.
6. Southern EM (1975) Detection of specific sequences among DNA fragments separated by gel electrophoresis. Journal of Molecular Biology 98: 503–517.

3

Northern Blotting: A Brief Overview

Prerna Pandey[1], Shivsanjeevi Sripathi[2]

[1]Dr Prerna Pandey, a biotechnologist with several years of wet lab research experience, has worked as a senior research fellow at International Center for Genetic Engineering and Biotechnology, New Delhi. Her field of research in PhD included isolation and molecular characterization of Geminiviruses, genome sequencing, gene annotation, as well as gene silencing using the RNA interference technology. She has published papers in peer-reviewed journals in the field and has a submitted a number of annotated Geminiviral genome sequences in the GenBank, including two novel ones. Prerna has also completed her editing and proof-reading courses from SfEP, UK and now works as a freelance scientific editor, writer and a subject matter expert. She has edited and co-authored several books on a range of topics in Life-Sciences. When Prerna is not busy with her assignments, she loves visiting historical places.

[2]ShivSanjeevi Sripathi completed his Masters in Biotechnology from Mumbai University in 2008. He was awarded for academic excellence in both his Bachelors and Masters for securing second rank in Mumbai University in 2006 and first rank in his college: Kishinchand chellaram College. For his Masters he secured first rank in his college KET's V.G.Vaze College. He qualified CSIR and NET and and TOEFL in September 2008. He then worked on a stem cell project at the Specialized Centre for Cell Based Therapy (SCCT),KEM Hospital at Mumbai on a project entitled,: Isolation & detection of stem cells from Human Umbilical cord/ amniotic membrane" following which he worked at Junior Research Fellow at Microbiology & Cell Biology Department, Indian Institute of Science, Bangalore from August 2009 to November 2011. The field of work involved cloning of cell wall genes and transcription factors in *E.coli & M.smegmatis*. As a writer, he has authored and co-authored 25 books with Delve and Arcler International Press. He loves to read and share on interesting aspects of life sciences in books. In his free time he loves to travel and explore ancient customs and traditions.

3.1 INTRODUCTION

One of the most popular techniques in specific RNA sequence detection is Northern blotting. At Stanford University, David Kemp, James Alwine and George Stark invented the technique of Northern blotting in 1977, and named it so as a reverence to Dr. Edwin Southern, who invented the Southern Blot. Over the years, the Southern Blotting was modified to accommodate the detection of many more target molecules. The variations in the Southern Blotting techniques were called the Northern Blotting, Eastern Blotting and Western Blotting. In a sample, the Northern blotting technique detects specific RNA sequences and hence is also referred to as the RNA Blot.

In Northern Blotting, gel electrophoresis is used to separate RNA from the sample on the basis of the molecular size of the strands. Once separated, the RNA fragments are exposed to a labelled DNA prove, after being fixed on a support membrane. If the labelled DNA Probe gets across the complementary RNA sequence, the probe will bind and the result can be visualized by appropriate methods (Rio et al., 2010; Cold Spring Harb Protoc; 2015).

3.2 THE PROCEDURE IN BRIEF

The following portray the vital steps in the process of the Northern blotting technique.

The first step agarose gel electrophoresis to separate the RNA sample. Since there are many tiny RNA fragments, the RNA molecules, rather than banding, actually form streaks (Rio et al., 2010; Cold Spring Harb Protoc; 2015).

A special blotting paper popularly made of nitrocellulose, is used to transfer the RNA molecules (Nylon membranes are also used for the RNA transfer). After transferring to the blotting paper, the pattern of separation of the RNA molecules that existed in the electrophoresis gel remains the same.

A labelled DNA probe is then introduced into the blotting paper. The RNA sequences in the membrane will bond with the DNA base strands in the probe to form a RNA-DNA hybrid. The probe is labelled with a radioactive tag or an enzyme, and will not be visible at this stage. However, at the end of the treatments in the following step the probe becomes visible. To make the substrate visible on an X-Ray film, an enzyme is used to make the colored product, as the probe is introduced to a substrate that is colorless. However, in the case of a radioactive probe, there is no need for an enzyme as the it is directly visible on an X-Ray film due to the radioactivity (ThermoFisher Scientific; 2019b).

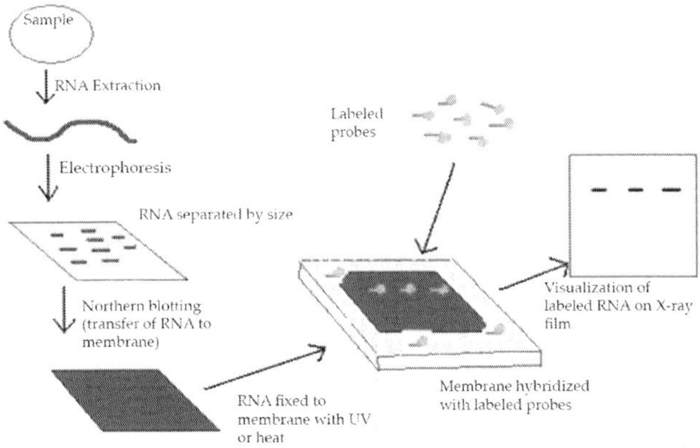

The schematics of Northern blotting (ThermoFisher Scientific; 2019b)

Since RNA is more likely to degrade, the electrophoresis of RNA comes with a prerequisite of heightened caution, when compared to the electrophoresis of DNA. RNase, which is found everywhere, must be absolutely eliminated from the apparatus and the reagents required to perform the electrophoresis of RNA, something that is best achieved with DEPC treatment (Diethyl Pyrocarbonate).

Denaturing of RNA pre-electrophoresis is very important, as single stranded RNA in its native form, has a propensity to form a secondary structure, and the RNA has to be maintained denatured during electrophoresis. Denaturing RNA is done by heating the sample along with formaldehyde and formamide to a temperature of fifty-five degrees centigrade. After that, during the process of electrophoresis, to prevent re-formation of secondary structures, additional formaldehyde is introduced, as required. However, it should be noted that ethidium bromide is not commonly advised to be added to the electrophoresis process, since, when compare with unstained RNA, this addition would reduce the hybridization signal all the time. Figure 1 elucidates a common Northern Hybridization protocol. As is the procedure in a Southern Blot, subsequent to electrophoresis, the RNA that has been separated is moved to a nylon membrane. This may be done by any of the three methods:

- In the commonly practiced ***upward capillary transfer***, the agarose gels and takes several hours (3–6) to overnight for a complete transfer (Figure 2a).

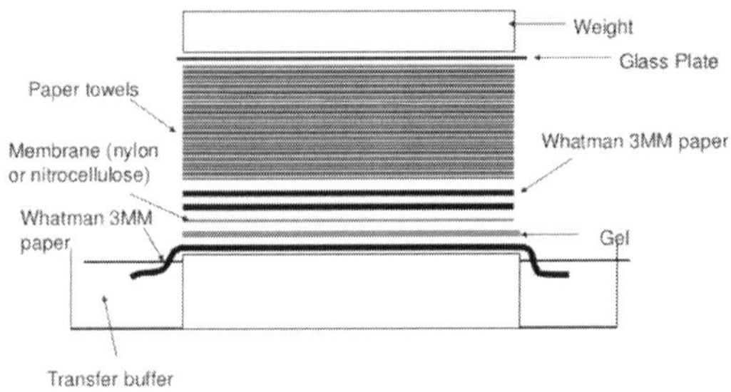

Figure 2a: Upward capillary transfer.

- The ***electroblotting transfer*** technique is used with polyacrylamide gels and the transfer usually gets done in less 60 min (Figure 2b).

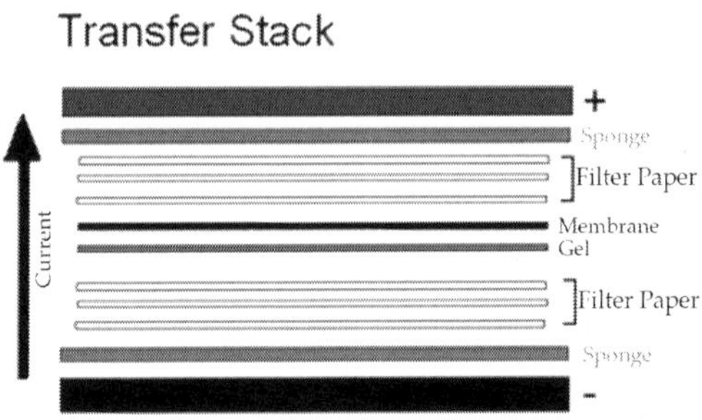

Following which, by baking the RNA at eighty degrees centigrade or by irradiating the RNA with ultraviolet light, fixation occurs.

3.3 APPLICATIONS OF THE NORTHERN BLOTTING TECHNIQUE

Studies on Gene Expression – In the cases of rejections of transplants, to study the gene expression, and in the cases of cancer-causing genes, to study their overexpression: In the course of a specific procedure of treatment, the study of an individual gene's expression profile amidst various environmental stress situations, levels of pathogen infection, stages of development, in organs and at tissue levels, Northern Blotting technique helps in their conduction (Mori, 1991; Liang, 1995; Baldwin, et al., 1999).

Further, in ***the detection of oncogene upregulation***, or in the tumor-suppressing neoplasia genes in the case of downregulation, the Northern blotting technique has been acclaimed to be very useful. When an increased expression of a specific gene observed through the Northern blot in the form of enhanced mRNA expression in the blot, further investigation is possible by sequencing with specific targets, to assess if it is an erstwhile known gene or unknown to the researchers (Utans, 1994).

The exact role of the gene under investigation can be better understood by studying its expression that is derived under the prescribed conditions. Moreover, there is the ability to understand whether the RNA bands on the membrane are sequential repetitive motifs or whether they are spliced species of the very same gene. This is possible because of the fact that the initial resolution of RNA is with its molecular weight, and by using a single probe the variation in each band on the membrane can be highlighted (Durand and Zukin, 1993; Gortner, et al., 1996).

Deletions and errors in processing of the transcript can also be brought to the forefront by the difference in the gene product fragment size. Online databases like the BlotBase publish outcomes of Northern blots. A detailed search of Northern Blots can be done in terms of paper reference, type of tissue, gene identifier or blot ID (Schlamp et al., 2008). By conducting the search, the expression level of the gene, corresponding tissue, species, the blot ID and the image (if provided) as well as the links to the original publication of the research can be obtained (Schlamp et al, 2008).

As a technique to diagnose various diseases: In the testing of patterns of gene expression in the case of human cancer, Northern blotting technique was the first mRNA detection technique used, as was with several other diseases (Streit et al., 2009). Due to various factors such as the presence of nonneoplastic tissues surrounding the malignant cells, dilution of the target mRNA levels occurs, the process itself being cumbersome and slow, the Northern blotting technique is generally restricted to the academic sectors and does not play a wide role in the assessment of actual human clinical samples.

The very first microRNA was identified using Northern blotting and it still prevails as the gold standard for expression analysis of microRNA (Várallyay et al., 2008). A high percentage electrophoresis gel is used to fractionate the tiny RNA that have been extracted from tissues and cells by electrophoresis. A nitrocellulose membrane is used to receive the miRNA transferred from the electrophoresis gel which are identified using a radioactive DNA probe that is labelled, or a fluorescent probe, that has complimentary characteristics to the target miRNA. The extracted RNA, by employing UV cross linking technique or by treating at specified temperature in an oven, are attached to the membrane.

The process of ***microRNA detection*** requires high sensitivity because microRNA are extremely small non-coded fragments, and due to their extremely small size and the fact that they occur in very few numbers, an oligonucleotide probe is required for their detection (Koscianska et al., 2011). The predicted miRNAs get validated by the Northern blotting technique by analyzing their levels of expression and molecular sizes.

Northern Blotting Technique is an excellent too ***to research on Gene Transcription products*** (Maamar et al., 2006), besides, setting up of a ***mRNA level standard***, to facilitate direct gene expression study viz. messenger RNA transcripts.

Other applications include:

Assessment of the ***messenger RNA size.***

RNA degradation studies.

RNA splicing study – the ability to identify transcripts that have been spliced alternatively.

Half-life study of RNA.

It is used ***to study the Internal Ribosomal Entry site (IRES)*** to eliminate the possibility of occurrence of RNA digestion as compared to the second cistron translation.

To generally check and for additional confirmation.

To ***screen the recombinants through the detection of mRNA*** that may arise from the transgene.

3.4 DRAWBACKS OF THE NORTHERN BLOTTING TECHNIQUE (TANIGUCHI ET AL., 2001; KAREN REUE, 1998)

Radioactivity is commonly used, which makes the usage and disposal difficult. Nevertheless, new non-radioactive detection techniques have been developed. Next, the duration of the Northern Blotting process from the preparation of the sample to the point of detection, is extremely long. Moreover, the expression quantitation and the expression of RNA data is negatively affected, in case there is degradation by RNAses.

Degradation of the sample is a major issue that can hamper quantitation as well as the quality of the samples, as the Northern blotting procedure has a very high sensitivity to RNA sample degradation. Further, RNA Stability and the rates of transcription of RNA are outside the purview of the Northern Blotting technique, as it is capable of measuring only RNA accumulation that is in a steady state.

Other techniques like Reverse Transcription Polymerase Chain Reaction and the nuclease protection assays are much more sensitive than the Northern blotting technique. However, by using positively charged nylon membranes and a highly specific antisense probe, the Northern blotting sensitivities can be enhanced. Besides, in the case of Northern Blotting, the volume of the sample sequence of RNA that is required is very huge, while in the case of other techniques, like real time Reverse Transcription Polymerase Chain Reaction, as low as a single copy can do the requisite.

A major technical issue is the use of multiple probes detection. The Northern Blotting technique is straightforward only as far as one probe is concerned. A second probe can be introduced only after the first one is removed completely, thereby complicating the process tremendously Potent chemicals are often used in the process of removing the first probe, and subsequently there is no relying on the validity of the data arising from the subsequent probes.

3.5 THE REVERSE NORTHERN BLOT TECHNIQUE

The reverse Northern blot is a procedure that researchers use sometimes (Kushwaha et al., 2015). Like the term denotes, in this procedure the substrate nucleic acid which gets to be fixed to the membrane is actually some isolated DNA fragments, and the RNA is radioactively labelled and forms the probe, upon extraction from a tissue. With the onset and gaining popularity of DNA arrays since the late nineties and early 2000s, the substrate is composed of DNA fragments, most of the time and the cellular RNA is used to form the probe. This onset of the reversing of the procedure, actually gave a hand to venture Northern blotting technique into the realm of profiling of gene expressions and therefore, despite not being common earlier, reverse northern blot can actually monitor almost all the gene expressions in an organism (Gilmartin et al., 2002).

3.6 THE NORTHWESTERN BLOT

A combination of the western blot and the northern blot techniques gave rise to the onset of the Northwestern blotting technique, also commonly referred to as the Northwestern assay, which finds a role in the field of molecular biology, and is popularly utilized to decipher RNA and protein interactions (Saminathan, 2014). In the case of the similar Western blotting technique, the specific protein, to be isolated, is transferred along with the other proteins, via gel electrophoresis over to a nitrocellulose base membrane. The band that has the desired protein, then features a colored precipitate that forms a cluster around it. The analytical Northern blotting technique is much similar to this, only difference being that the gene expression study of RNA and mRNA is possible and not of a protein. The Northwestern blot is an amalgam of these two techniques and features specifically the labelled RNA interaction with proteins as they remain immobilized in a nitrocellulose membrane that is quite similar to the Northern blot (Zang and Lin, 2016)

An important part of the Northwestern Blot is the procedure for separating the RNA from the proteins that bind them, and the parameters of this separation is described by the state of charge in the electrophoresis gel and on the size of the molecules. To study more than one sample simultaneously, each separate sample can be placed on the polyacrylamide gel, usually sodium dodecyl sulfate. Upon completion of the electrophoresis, the gel as well as the proteins that had bound to the RNA get transferred to the nitrocellulose membrane used for the transfer.[7]

After blocking, an incubation session at room temperature is initiated with the application of a specified competitor RNA. With the introduction of the competitor RNA, bonding occurs with that RNA to the proteins that are there on the sample in the blot. Post incubation, the intent is now to dilute the RNA in the solution and therefore the blot is washed for at least three to five minutes. A washed blot is then developed and visualized.

The results from the autoradiography developed blot can be studied to ascertain the projected size and the density of the proteins that have bound the RNA and hence can provide the motivation for further research of the said protein. Upon determining the size of the said protein, even as an approximate, the chromatography machine can be used to run the original sample with the exclusion of the known size, and therefore separate the proteins (Gray et al., 2008). Once the protein is separated it can be digested with the introduction of trypsin, and the peptide sequences can be created using mass Spectrometry, and the specific protein can be found.

A definite benefit of Northwestern botting technique is the identification of the molecular weight of the specified proteins as well as fast detection of the said proteins that bind the RNA. Another palpable advantage is that it is inexpensive. The Northwestern blot is usually the first step in a research procedure and since the molecular weight can be ascertained rather quickly, further study is possible and other techniques for purification like chromatography and so on can also be used. The expression libraries of the cognate ligands are also being built with the help of the Northwestern blot technique.

One of the chief drawbacks of the Northwestern Blot is that RNA-protein interaction with RNA that have poor binding characteristics, become undetectable.[15] The duration of three to five hours of the Northwestern Blot is also another drawback. Any error in the performance of the Blot will leave enough background to make the detection of the proteins impossible. Moreover, renaturing of the proteins is a must after the separation and movement to the nitrocellulose paper. And the final drawback is that the proteins that comigrate to the gel matrix have to be comprised of a single polypeptide or two subunits (Allen, 2001).

3.7 SUMMARY

The Northern Blot technique - a classical way of studying the presence of an RNA sequence in a sample. It is an inexpensive procedure that can readily and easily be performed in the lab. Moreover, the issue of sensitivity in the blotting procedure has also improved a lot with the technological leapfrogging in the development of the hybridization membranes and the buffer composition. Similarly, huge advancements in technologies of the Polymerase Chain Reaction, have made it possible for precision identification as well as excellent quantitation of the RNA in a simple, efficient and quick manner.

3.8 REFERENCES

1. Rio DC, Ares M, Hannon GJ, Nilsen TW. 2010. Polyacrylamide gel electrophoresis of RNA. Cold Spring Harb Protoc, 10.1101/pdb.prot5444.
2. Cold Spring Harb Protoc. 2015. Northern Blots: Capillary Transfer of RNA from Agarose Gels and Filter Hybridization Using Standard Stringency Conditions, doi:10.1101/pdb.prot081018
3. ThermoFisher Scientific. 2019b. The basics: Northern analysis. [Cited 2019 Feb 15]. Available from: https://www.thermofisher.com/ca/en/home/references/ambion-tech-support/northern-analysis/general-articles/the-basics-northern-analysis.html
4. Schlamp K, Weinmann A, Krupp M, Maass T, Galle P, Teufel A. 2008. BlotBase: a northern blot database. Gene, 427(1-2):47-50. doi: 10.1016/j.gene.2008.08.026.
5. Streit S, Michalski CW, Erkan M, Kleeff J, Friess H. 2009. Northern blot analysis for detection and quantification of RNA in pancreatic cancer cells and tissues. ,.4(1):37-43. doi: 10.1038/nprot.2008.216.
6. Várallyay E, Burgyán J, Havelda Z. 2008. MicroRNA detection by northern blotting using locked nucleic acid probes. Nat Protoc, 3(2):190-6. doi: 10.1038/nprot.2007.528.
7. Koscianska E, Starega-Roslan J, Sznajder LJ, Olejniczak M, Galka-Marciniak P, Krzyzosiak WJ. 2011. Northern blotting analysis of microRNAs, their precursors and RNA interference triggers. BMC Mol Biol, 12:14. Published 2011 Apr 11. doi:10.1186/1471-2199-12-14
8. Maamar H, Abdou L, Boileau C, Valette O, Tardif C. 2006. Transcriptional Analysis of the *cip-cel* Gene Cluster from *Clostridium cellulolyticum*. 2006. Journal of Bacteriology, 188(7): 2614-2624.
9. Taniguchi M, Miura K, Iwao H, Yamanaka S. 2001. Quantitative Assessment of DNA Microarrays – Comparison with Northern Blot Analysis. Genomics, 71 (1): 34–39.
10. Karen Reue. 1998. mRNA Quantitation Techniques: Considerations for Experimental Design and Application. The Journal of Nutrition, 128(11): 2038–2044.
11. Kushwaha, Nirbhay & Sahu, Pranav & Prasad, Manoj & Chakraborty, Supriya. 2015. Chilli leaf curl virus infection highlights the differential expression of genes involved in protein homeostasis and defense in

resistant chilli plants. Applied microbiology and biotechnology, 99. 10.1007/s00253-015-6415-6.

12. Philip M. Gilmartin, P. M. Gilmartin, Dr. Chris Bowler. Molecular Plant Biology: A Practical Approach, Volume 2. Oxford University Press, 2002.
13. Saminathan T. 2014. Northwestern Blot of Protein-RNA Interaction from Young Rice Panicles. Retrieved 26 March 2014.
14. Zang S, Lin RJ. 2016. Northwestern Blot Analysis: Detecting RNA-Protein Interaction After Gel Separation of Protein Mixture. Methods Mol Biol, 1421:111-25. doi: 10.1007/978-1-4939-3591-8_10.
15. Gary HP, Heuvel JPV; Peters JM. Regulation of Gene Expression: Molecular Mechanisms. Springer. p. 129. *ISBN 9781597452281*.
16. Allen WN. 2001. Ribonucleases, Part A: Functional Roles and Mechanisms of Action: Functional Roles and Mechanisms of Action. Academic Press. p. 409. ISBN 9780080496917.

4

Eastern Blotting

Shivsanjeevi Sripathi

ShivSanjeevi Sripathi completed his Masters in Biotechnology from Mumbai University in 2008. He was awarded for academic excellence in both his Bachelors and Masters for securing second rank in Mumbai University in 2006 and first rank in his college: Kishinchand chellaram College. For his Masters he secured first rank in his college KET's V.G.Vaze College. He qualified CSIR and NET and and TOEFL in September 2008. He then worked on a stem cell project at the Specialized Centre for Cell Based Therapy (SCCT),KEM Hospital at Mumbai on a project entitled, : Isolation & detection of stem cells from Human Umbilical cord/ amniotic membrane" following which he worked at Junior Research Fellow at Microbiology & Cell Biology Department, Indian Institute of Science, Bangalore from August 2009 to November 2011. The field of work involved cloning of cell wall genes and transcription factors in *E.coli* & *M.smegmatis*. As a writer, he has authored and co-authored 25 books with Delve and Arcler International Press. He loves to read and share on interesting aspects of life sciences in books. In his free time he loves to travel and explore ancient customs and traditions.

4.1 OUTLINE

This technique is described for the analysis of post-translational modifications (PTM) on proteins: mostly carbohydrate epitopes or lipids or phosphomoieties. It is usually described as an extension of Western blotting. Following the blotting of proteins on SDS-PAGE followed by transfer to a nitrocellulose membrane, probes to detect the modifications are applied.

Figure 4.1 summarizes the definitions and concepts of Eastern blotting:

(1982) The term *Eastern blotting* was specifically rejected by two separate groups: Reinhart and Malamud referred to a protein blot of a native gel as a *native blot*; Peferoen et al., opted to refer to their method of drawing SDS-gel separated proteins onto nitrocellulose using a vacuum as *Vacuum blotting*.

(1984) *Middle Eastern blotting* has been described as a blot of polyA RNA (resolved by agarose) which is then immobilized. The immobilized RNA is then probed using DNA.

(1996) *Eastern-Western blot* was first used by Bogdanov et al. The method involved blotting of phospholipids on PVDF or nitrocellulose membrane prior to transfer of proteins onto the same nitrocellulose membrane by conventional Western blotting and probing with conformation specific antibodies.
This method is based on earlier work by Taki et al. in 1994, which they originally dubbed *TLC blotting*, and was based on a similar method introduced by Towbin in 1984.

(2000) *Far-Eastern blotting* seems to have been first named in 2000 by Ishikawa & Taki. The method is described more fully in the article on Far-Eastern blotting, but is based on antibody or lectin staining of lipids transferred to PVDF membranes.

(2001) *Eastern blotting* was described as a technique for detecting glycoconjugates generated by blotting BSA onto PVDF membranes, followed by periodate treatment. The oxidized protein is then treated with a complex mixture, generating a new conjugate on the membrane.
The membrane is then probed with antibodies for epitopes of interest. This method has also been discussed in later work by the same group.
The method is essentially Far-Eastern blotting.

(2002) *Eastern blot* has also been used to describe an immunoblot performed on proteins blotted to a PVDF membrane from a PAGE gel run with opposite polarity.

(2005) *Eastern blot* has been used to describe a blot of proteins on PVDF membrane where the
probe is an aptamer rather than an antibody.
This could be seen as similar to a Southern blot, however the interaction is between a DNA molecule(the aptamer) and a protein, rather than two DNA molecules.
The method is similar to South-western blotting.

(2006) *Eastern blotting* has been used to refer to the detection of fusion proteins through complementation.
The name is based on the use of an enzyme activator (EA) as part of the detection.

(2009) *Eastern blot* has been used to describe a blot of proteins on nitrocellulose membrane where the probe is an aptamer rather than an antibody.
The method is similar to South-western blotting.

(2011) The study used the term Eastern blotting to describe detection of glycoproteins with lectins such as concanavalin A.
There is clearly no single accepted definition of the term.

The process of post-translational modifications (PTMs) occurs once nascent proteins are synthesized and cover a broad range of reactions such as phosphorylation, alkylation, ADP-Ribosylation, acetylation, formylation, biotinylation, etc. they become significant in the case of transport of secretory proteins and diseases.

(World Heritage Encyclopedia)

4.2 CASE STUDY 1: EASTERN BLOTTING FOR A HERB VS. TLC (THIN LAYER CHROMATOGRAPHY)

The herb *Scutellaria baicalensis* Georgi (Labiatae) is the source of *Scutellariae radix* (*S. radix*)-the dried root of *S. baicalensis*. This root is the source of important drugs used in traditional Chinese medicines and Kampo medicines. These drug molecules include baicalin (BI) or baicalein (BE), flavonoids present in the root that scavenge free radicals to show anti-inflammatory, anti-allergic and anti-viral effects. In the case of oxidative burst against plant disease and insect attack, BI has been shown as the major player.

A team led by Osamu Morinaga reported the use of Eastern blotting to analyze baicalin using anti-BI MAb or monoclonal antibodies. For immunostaining, the BI molecule was separated into two functional parts, the epitope part (mainly aglycone) and the sugar part. Oxidation of the latter yields aldehyde groups which react with amino groups of lysine or arginine to bind to the adsorbent membrane (polyethersulfone (PES) membrane). The mAb anti-BI MAb binds the epitope.

Figure 4.2 shows the scheme of immunostaining (Morinaga et al, 2012):

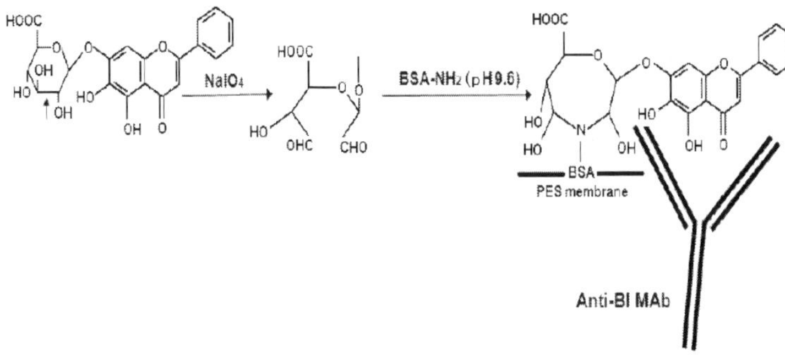

The system was so designed that the process of immunostaining involved the blotting of BI from the TLC plate to the PES membrane. This was followed by treatment with NaIO4 solution, followed by the addition of BSA to augment the fixation process. When compared to TLC (FeCl3 stained), the sensitivity of the developed eastern blotting was similar to TLC plus FeCl3 staining. The detection of BI was 1μg on the eastern blotting membrane. Figure 4.3 below shows the eastern blotting sensitivity (Morinaga et al, 2012):

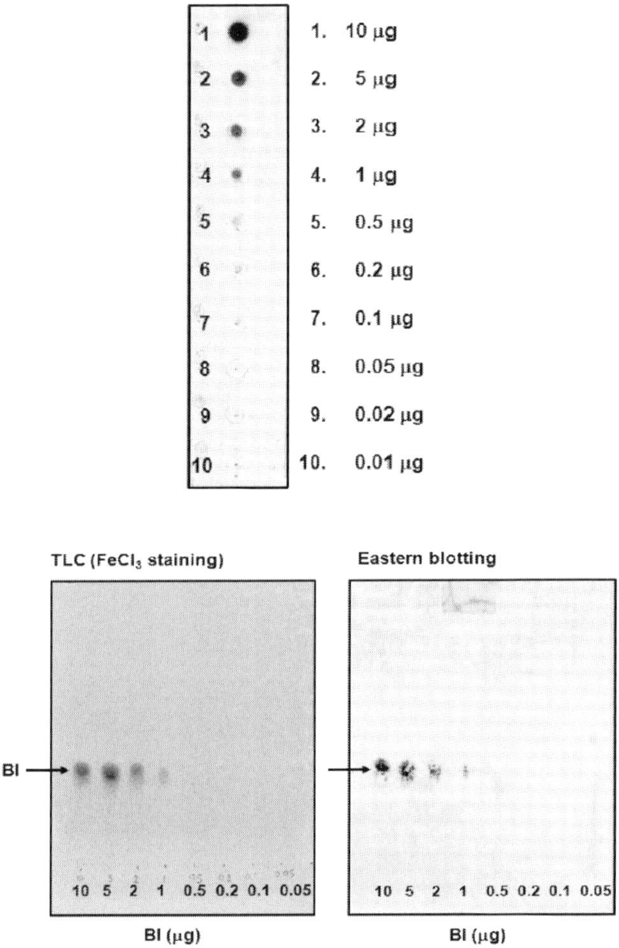

The eastern blotting technique was used for analyzing *S. radix* samples and Kampo medicines. *S. radix* bands appeared in BI containing extracts except Lane 6 in figure 4.4 below which is shakuyakukanzoto extract lacking BI. Quantitatively, the analysis of anti-BI MAb revealed that quantitatively eastern blotting.

Furthermore, the concentration of BI was quantitatively analyzed by competitive ELISA using anti-BI MAb (Table 1). Collectively, this established eastern blotting technique is more useful than the TLC analysis for the specific detection of BI in medicines that contain a mix of compounds rather than TLC.

Figure 4.4 below shows the detection of BI in several medicines: 1. Baicalin standard, 2. *S. radix* main root, 3. *S. radix* A, 4. *S. radix* B, 5. *S. radix* C, 6. Shakuyakukanzoto, 7. Shosaikoto, 8. Saikokaryukotsuboreito (Morinaga et al, 2012):

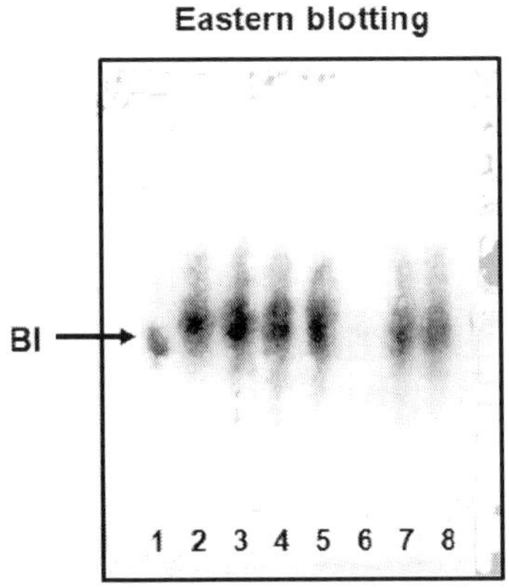

The application of eastern blotting was extended to observing the distribution of BI in the fresh root of *S. baicalensis*. Following the inactivation of endogenous peroxidases using methanol, the root was pressed against a PES membrane followed by immunostaining. BI was found to be more in the cortex (A in figure 4.5) than the xylem (B in Figure 4.5): (Morinaga et al, 2012):

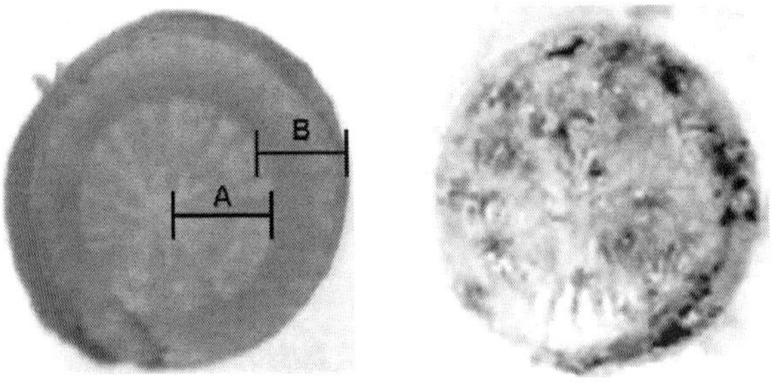

It was concluded that the visualization of a molecule with low molecular weight was facilitated by eastern blotting. Briefly, the process involved the application of the test compound on a TLC plate followed by development and blotting. The components were then transferred on to a PES membrane that was then subjected to immunostaining. This can allow the identification of a test molecule from a large mixture of molecules as located in a crude drug extract (Morinaga et al, 2012).

4.3 CASE STUDY 2: EASTERN BLOTTING FOR BIOACTIVE COMPOUNDS

Several herbal medicines used in Japan and traditional Chinese medicine find the application of licorice as a pain-reliever, expectorant and antitussive. A major component behind the activities is due to glycyrrhizin (GC). An example is Figure 4.6: *Glycyrrhiza uralensis* (a) flowers (b) fruits and seeds, and (c) licorice roots along with the structures of glycyrrhizin: (Yukihiro, 2013)

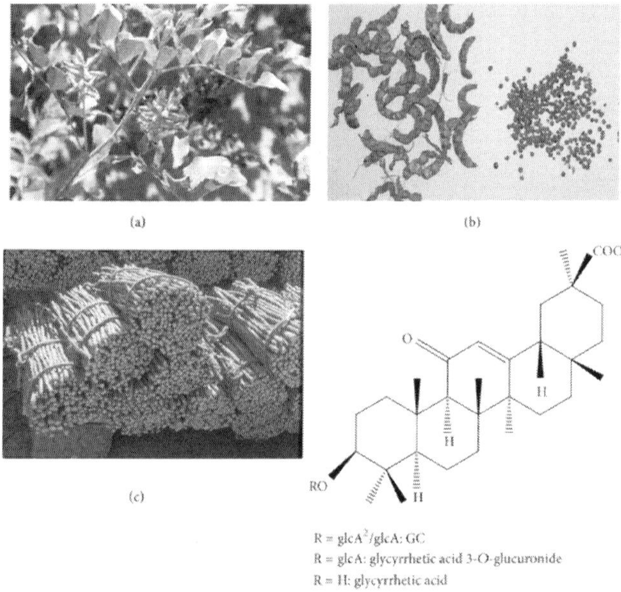

The standardization of herbal medicines requires the precise determination of the phytochemical constituents. Several techniques have been used such as chromatography, nuclear magnetic resonance and immunoassays using mAbs.

$NaIO_4$ solution was added to glycyrrhizin (GC) to release the aldehyde of the sugar to be combined with a carrier protein to prepare antiglycyrrhizin

MAb. This was followed by fingerprinting using Eastern blotting; a polyvinylidene difluoride (PVDF) membrane covered the TLC plate followed by treatment with anti-GC MAb and then peroxidase-labeled goat anti-mouse IgG MAb and subsequent substrate addition.

Figure 4.7 below shows the scheme of the eastern blotting (Yukihiro, 2013):

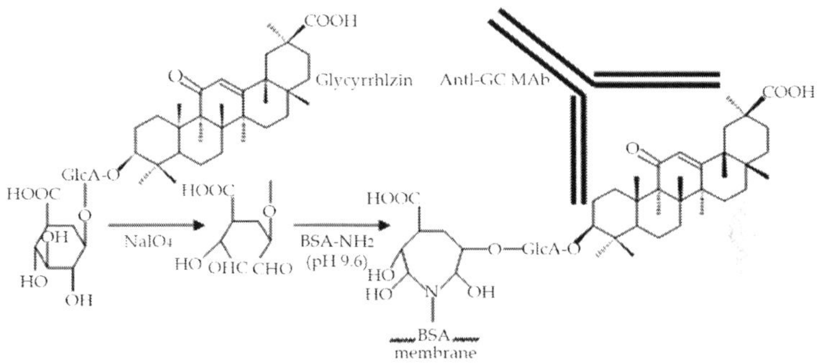

Figure 4.8 below shows the results of the eastern blot for GC in medicines; the sensitivity of the blotting was more than GC detection using H2SO4 staining on TLC (Figure 4.8 a) and no band for GC was seen in the Da Chaihu Tang prescription that lacks licorice (Lane 8) showing the specificity of eastern blotting to detect bioactive molecules (Figure 4.8 b): (Yukihiro, 2013)

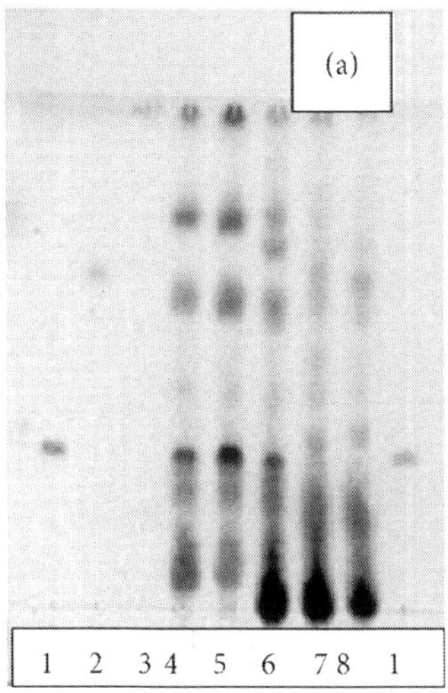

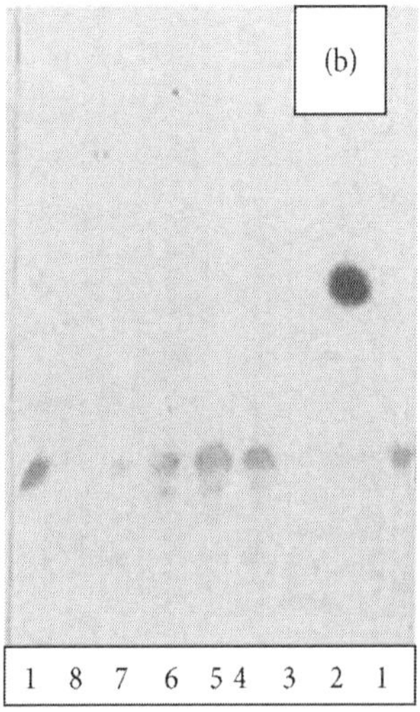

The use of eastern blotting in detecting GC in rat serum was checked; after injection of glycyrrhizin in rat serum, TLC stained by H2SO4 and eastern blotting were used for analyzing the chromatographic fingerprinting of GC as shown below in Figure 4.9 (Yukihiro, 2013):

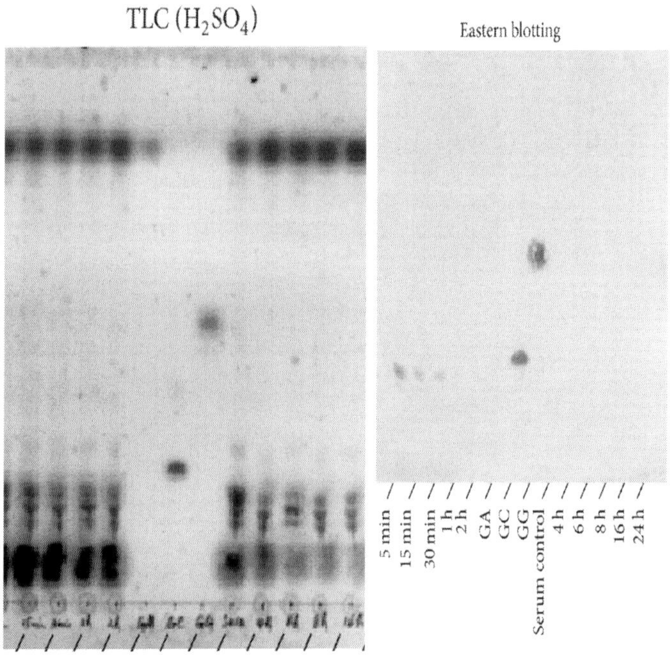

Despite the presence of many bands on TLC, GC could not be detected while the band of GC was seen in eastern blotting.

The application of eastern blotting was extended to examine the immunolocalization of GC in the root slices of licorice. GC was found more in the phloem than the xylem. The patterns were different for different *Glycyrrhiza* species like *G. uralensis*, *G. glabra,* and *G. inflata* stained by eastern blotting thus aiding standardization. Figure 4.9 below shows immunostaining using eastern blotting (Yukihiro, 2013):

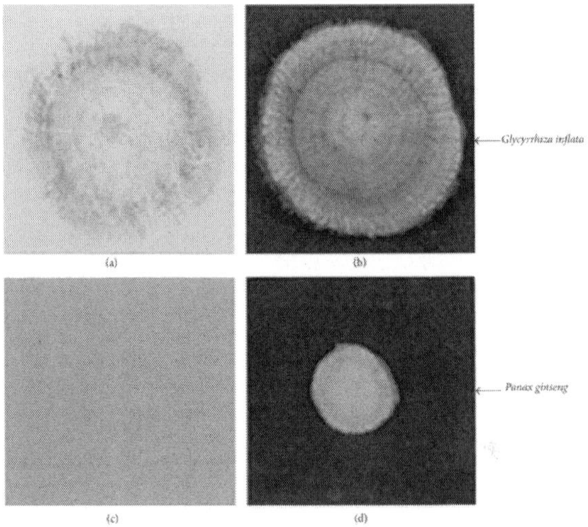

Thus, the approach termed as eastern blotting finds its place in the detection of a specific bioactive molecule in a mixture. This can assist the standardization of herbal medicines.

(Yukihiro, 2013)

4.4 REFERENCES

1. Eastern blotting, World Heritage Encyclopedia, Article ID: WHEBN0021890398.
2. Morinaga O, Mukae R, Uto T, et al. Development of Eastern Blotting Technique for Analysis of Baicalin Using Anti-Baicalin Monoclonal Antibody. Antibodies. 2012; 1:284-293. doi:10.3390/ antib1030284.
3. Yukihiro S. Standardization of Licorice and TCM Formulations Using Eastern Blot Fingerprinting Analysis. J Chem. 2013. 10.1155/2013/573070.

5

Misleading Westerns: Common Quantification Mistakes in Western Blot Densitometry and Proposed Corrective Measures

Trent A. J. Butler,[1,2] Jonathan W. Paul,[1,2] Eng-Cheng Chan,[1,2] Roger Smith,[1,2,3] and Jorge M. Tolosa[1,2]

[1]Mothers and Babies Research Centre, School of Medicine and Public Health, Faculty of Health and Medicine, University of Newcastle, Callaghan, NSW, 2308, Australia
[2]Hunter Medical Research Institute, 1 Kookaburra Circuit, New Lambton Heights, NSW, 2305, Australia
[3]John Hunter Hospital, New Lambton Heights, NSW, 2305, Australia

ABSTRACT

Densitometry data generated for Western blots are commonly used to compare protein abundance between samples. In the last decade, it has become apparent that assumptions underpinning these comparisons are often violated in studies reporting Western blot data in the literature. These violations can lead to erroneous interpretations of data and may contribute

Citation: Misleading Westerns: Common Quantification Mistakes in Western Blot Densitometry and Proposed Corrective Measures. Trent A. J. Butler, Jonathan W. Paul, Eng-Cheng Chan, Roger Smith, and Jorge M. Tolosa. Hindawi BioMed Research International Volume 2019, Article ID 5214821, 15 pages https://doi.org/10.1155/2019/5214821

to poor reproducibility of research. We assessed the reliability of Western blot data obtained to study human myometrial tissue proteins. We ran dilution series of protein lysates to explore the linearity of densitometry data. Proteins analysed included αSMA, HSP27, ERK1/2, and GAPDH. While ideal densitometry data are directly proportional to protein abundance, our data confirm that densitometry data often deviate from this ideal, in which case they can fit nonproportional linear or hyperbolic mathematical models and can reach saturation. Nonlinear densitometry data were observed when Western blots were detected using infrared fluorescence or chemiluminescence, and under different SDS-PAGE conditions. We confirm that ghosting artefacts associated with overabundance of proteins of interest in Western blots can skew findings. We also confirm that when data to be normalised are not directly proportional to protein abundance, it is a mistake to use the normalisation technique of dividing densitometry data from the protein-of-interest with densitometry data from loading control protein(s), as this can cause the normalised data to be unusable for making comparisons. Using spiked proteins in a way that allowed us to control the total protein amount per lane, while only changing the amount of spiked proteins, we confirm that nonlinearity and saturation of densitometry data, and errors introduced from normalisation processes, can occur in routine assays that compare equal amounts of lysate. These findings apply to all Western blot studies, and we highlight quality control checks that should be performed to make Western blot data more quantitative.

INTRODUCTION

Western blotting is routinely used to detect proteins and their posttranslational modifications (PTM) in biological samples. The development and widespread uptake of modern imaging devices that capture digital images of Western blots, visualised through chemiluminescence or fluorescence detection, have resulted in the assumption of quantitative Western blotting in the literature. Each pixel in these digital images is assigned an intensity value that is related to the number of photons detected by the corresponding pixel in the sensor until it reaches saturation [1]. Non-detector saturated images are used for quantitation, and protein/PTM abundance is most often measured using an optical density (O.D.) algorithm, which calculates O.D. values from the background-corrected band intensity and band area of the protein/PTM of interest. All dedicated Western blot analysis software from major suppliers provides the ability to measure O.D. with minor vendor-vendor variations in implementation.

Ideal O.D. data are proportional to protein/PTM abundance and are therefore best modelled with linear regression through the origin (henceforth referred to as the directly proportional or proportional linear model, equation y = mx) [2–5]. It is commonly assumed that this model fits O.D. data [5], and only rarely are experimental data supporting this assumption included in published studies. Recent studies have, however, shown that O.D. data can violate this assumption and in these cases is better modelled with nonproportional linear functions (equation y = mx + b) or nonlinear hyperbolic functions [2, 4–6]. Identification of the appropriate model for O.D. data is critical because analysis of data using an inappropriate model results in incorrect estimation of the magnitude of any differences in protein/PTM abundance between samples [2, 4, 5, 7]. In extreme cases, Western blot data can become saturated (independent of saturation of the detection system), after which further increases in protein abundance cannot be detected and/or measured [2–21]. It is inappropriate to quantitatively analyse saturated Western blot data as differences in protein/PTM abundance between samples can be missed, resulting in false-negative findings [3, 4, 8, 12, 13].

Most quantitative Western blot studies normalise levels of target protein(s)/PTM(s) to levels of loading control protein(s) that do not change in abundance between comparisons [7, 9–12, 17, 18, 20, 22]. This normalisation method is used to correct for differences in protein abundance that are not relevant to the biological question being addressed [7, 9–12, 17, 18, 20, 22]. Normalisation is routinely performed by dividing the O.D. value of the target protein/PTM by the O.D. value of loading control protein(s) that were detected in the same sample [6, 10, 12, 13] and ideally were run in the same gel lane and detected in the same Western membrane [23]. This ratio (target/loading control(s)) is used for comparisons and any difference between samples is believed to correspond to the difference in target protein/PTM abundance. It is not widely recognised, however, that this normalisation strategy is only valid when O.D. data fit directly proportional models [4–6, 24]. Considering that in the best case scenario O.D. data only conform to directly proportional models over limited dilution ranges of lysate [2–21, 25], it is possible that many Western blot studies using this normalisation approach have been compromised by incorrect assumptions.

Western blots are among the most prevalent experimental tools used to study proteins in reproductive biology, which includes studies on myometrial smooth muscle function. In this field it is standard practice to normalise O.D. data of target proteins to loading control protein(s), such as alpha-smooth muscle actin (αSMA) [26–28], β-actin [29], glyceraldehyde 3-phosphate dehydrogenase (GAPDH) [30], and α-tubulin [31]. It is also common for PTM specific data to be normalised to total levels of the target protein [28, 32, 33]. We performed a random search of over 100 PubMed-indexed papers published between 2005 and 2017 that investigated myometrial function using quantitative Western blotting and found that only one [30] had presented Western blot validation data in either the main text or supplementary material.

Our study aimed to test whether quantitative Western blotting experiments, using untested assumptions of proportionality for normalisation, can be misleading and affect the study of human myometrial tissue proteins. Similar mistakes can befall all Western blot studies regardless of the sample type [2–21, 25]. While uncertain antibody specificity threatens the validity of any Western blot assay [2, 3, 20, 34–37], a Western blot performed using an antibody that does not recognise the correct target can only be fixed by replacing it with one that does. We therefore focus on possible errors that may occur during the quantification steps of Western blotting.

As part of these studies we comprehensively compared Western blot detection by chemiluminescence and infrared fluorescence and we provide an example of a quantitative Western blotting analysis using a proper workflow.

We hope that increased knowledge of the limitations of Western blot quantification will allow scientists to better use this technology to produce meaningful data and consequently lead to the routine inclusion of validation data in future publications.

MATERIALS AND METHODS

Protein Extraction and Preparation

These studies were approved by the Hunter and New England Area Human Research Ethics Committee, adhering to guidelines of the University of Newcastle and John Hunter Hospital, Newcastle, Australia (02/06/12/3.13). Myometrial tissue biopsies were collected during caesarean section from term nonlabouring and term labouring women who had provided informed written consent. These samples were snap frozen in liquid nitrogen (LN_2) and stored at -80°C. When processed, the samples were always stored on dry ice between steps until they were immersed in protein lysis buffer. Tissues were crushed in a metal mortar and pestle. Samples were placed into the mortar in a small pool of LN_2 and crushed by hitting the pestle with a mallet immediately after the LN_2 had evaporated. Following crushing, samples were immersed in LN_2 then transferred to individual CK28-R 1.5 mL Precellys tubes (Cat. No. KT03961-1-007.2, Bertin Technologies). The mortar and pestle was cleaned with 70% ethanol and precooled with LN_2 for each sample. When all samples were crushed, they were placed on ice and lysed in 1 mL of 2D extraction buffer (8 M Urea, 2 M Thiourea, 4% CHAPS) or 1 mL of SDS extraction buffer (2% SDS, 50 mM Tris pH 6.8, 5 mM EDTA). cOmplete mini-protease inhibitor cocktail tablets (Cat. No. 4693124001, Roche) and PhosSTOP phosphatase inhibitor tablets (Cat. No. 4906837001, Roche) were used in both lysis buffers. After addition of lysis buffer, the samples were immediately homogenised in a Precellys 24 homogeniser (Bertin Technologies) at 5000 rpm using 2 × 30 sec homogenisation intervals with a 20 sec break between each homogenisation. Samples were then stored on ice for 2 min to prevent sample warming and the Precellys homogenisation step was repeated twice more. Samples were then spun for 10 min at 16000*g* at 4°C to pellet insoluble cellular debris. Supernatant was extracted, transferred to fresh tubes, and stored at -80°C. The amount of protein in each sample was estimated using protein quantification assays run according to manufacturer's instructions. 2D extracts were quantified using the 2-D Quant Kit (Cat. No. 80-6483-56, GE Life Sciences). SDS samples were quantified using the BCA Protein Assay Kit (Cat. No. 23227, Thermo Fisher Scientific). Measurements were performed in a flat-bottomed 96-well plate at 480 nm wavelength for the 2-D Quant Kit and 562 nm wavelength for BCA Protein Assay Kit using a SPECTROstar Nano plate reader (BMG LABTECH). Pooled samples were created by mixing equal protein amounts (μg of total protein) of individual samples that were extracted in the same lysis buffer. The protein concentration in pooled samples was reestimated using the appropriate quantification kit listed above for 2D or SDS buffer. Each individual sample as well as pooled samples was aliquoted and stored at -80°C.

Spiking of Myometrial Homogenates with Recombinant Proteins

Protein spiking experiments used pooled human myometrium homogenates made from 3 nonlabouring tissues that were extracted in 2D lysis buffer. Extraction of these samples, pooling, and protein quantification were performed identically to the 2D extractions described above. This pooled sample was diluted to 2 mg/mL in 1 mL of 2D lysis buffer. 900 μL (1800 μg) of this sample was used to reconstitute 100 μg of purified recombinant ENPP1 (Cat. No. Ab167943, Abcam) and then 100 μg of Fam3a (Cat. No. Ab167946, Abcam) and then made to a final volume of 1 mL. A subsequent total protein quantification assay (2-D Quant Kit run according to manufacturer's instructions) confirmed that the protein content was still approximately 2 mg/mL. This method of protein spiking created a sample in which each spiked protein constituted approximately 10% of the total amount of protein in the lysate. This sample was then serially diluted 1:1 with the 2 mg/mL unspiked pooled lysate to create eight 2-fold serial dilutions that contained decreasing amounts of the spiked proteins at similar concentrations of total protein. These samples were then aliquoted and stored at -80°C, and an individual aliquot was used for each membrane. Protein separation and Western blotting were performed using the same equipment as the SDS samples described below. Western blot signals for recombinant ENPP1 and Fam3a were simultaneously imaged in the same Western blot membranes using an anti-6x-His-tag antibody that recognised the 6x-His-tag present on both proteins.

SDS-Polyacrylamide Gel Electrophoresis and Membrane Transfer

Stock aliquots of pooled myometrial tissue homogenates were defrosted on ice, warmed to room temperature, and then vortexed to ensure complete suspension of proteins. Samples were checked by eye to ensure that precipitates were not present. If precipitates were seen, the samples were left for 1 min at room temperature and then revortexed. This was repeated until samples became translucent. Samples were then stored on ice wherever possible, taking care to avoid the formation of precipitants. Each dilution of lysate was prepared as a stock sample with a volume of at least 10 μL and an independently prepared stock sample was used for each membrane in which the same protein(s) were detected. Samples loaded into gels were made up to the same final volume using the necessary amounts of stock sample, lithium dodecyl sulphate (LDS) sample buffer, reducing agent, and lysis buffer. Each sample was heated at 70°C for 10 min, then centrifuged for 15 secs on a Heraeus Pico microfuge set to 16000*g*, and loaded into their respective gels. Whether each dilution series was serial or independent dilutions is listed in the figure legends. Novex Sharp Prestained Protein Standard (Cat. No. LC5800, Thermo Fisher Scientific), MagicMark XP (Cat. No. LC5602, Thermo Fisher Scientific), and SeeBlue Plus2 Prestained Protein Standard (Cat. No. LC5925, Thermo Fisher Scientific) were used for molecular weight sizing.

Separation and Protein Transfer to Membranes of 2D Samples That Did Not Contain Recombinant Proteins

All 2D samples that were not spiked with recombinant proteins were run on 12-well 1 mm thick 4-12% NuPAGE Bis-Tris gels (Cat. No. NP0322, Thermo Fisher Scientific). Each sample loaded consisted of lysate diluted in 2D lysis buffer and the appropriate amount of NuPAGE LDS Sample Buffer (Cat. No. NP0008, Thermo Fisher Scientific), and NuPAGE Sample Reducing Agent (Cat. No. NP0009, Thermo Fisher Scientific). Protein separation was performed immediately following gel loading at 200 V, 120 mA, and 25 W for 55 min in NuPAGE MOPS SDS Running Buffer (Cat. No. NP0001-02, Thermo Fisher Scientific). Proteins were then transferred onto Optitran BA-S 85 Reinforced Nitrocellulose (Cat. No. 10439196, GE Life Sciences) or Immobilon-PSQ Polyvinylidene Difluoride (PVDF) membrane (Cat. No. ISEQ00010, EMD Millipore) in 1× NuPAGE Transfer Buffer (Cat. No. NP0006-1, Thermo Fisher Scientific) containing 10% methanol. PVDF membranes were prewet in 100% methanol for 1 min before transfer. Proteins were transferred at 25 V, 120 mA, and 17 W for 1 h and 10 min. SDS-PAGE and transfer steps were performed in an XCell SureLock Mini Gel Tank (Thermo Fisher Scientific) on a Power Zoom-dual power pack or Power Ease 500 power pack (Thermo Fisher Scientific).

Separation and Protein Transfer to Membranes for SDS Samples, and 2D Samples Containing Recombinant Proteins

All SDS samples were run on 10-well 1 mm thick 4-12% Bolt Bis-Tris Plus gels (Cat. No. NW04120, Thermo Fisher Scientific). Each sample loaded into each gel consisted of lysate diluted in SDS lysis buffer and the appropriate amount of Bolt LDS Sample Buffer (Cat. No. B0007, Thermo Fisher Scientific) and Bolt Reducing Agent (Cat. No. B0009, Thermo Fisher Scientific). 2D samples containing recombinant proteins shown in Figure 4 were run on the same type of gels and transferred to the same membrane type using identical conditions. After loading the gels, proteins were separated at 165 V and 125 mA for 45 min in 1× Bolt MOPS SDS Running Buffer (Cat. No. B000102, Thermo Fisher Scientific), and then proteins were transferred to Odyssey nitrocellulose membranes (Cat. No. 926-31092, LI-COR Bioscience) at 10 V and 160 mA for 1 h using 1× Bolt Transfer Buffer (Cat. No.BT00061, Thermo Fisher Scientific) containing 10% methanol. All SDS-PAGE and Western blot transfer of these protein samples were performed in Bolt mini-gel tanks (Thermo Fisher Scientific) using PowerEase 90 W power packs (Thermo Fisher Scientific). To make the example Western blot shown in Figure 5, two samples containing recombinant ENPP1 and Fam3a proteins were diluted 1 in 10 with the labelling buffer supplied in an Amersham QuickStain kit (Cat. No. RPN4000, GE Life Sciences) to a protein concentration of 0.2 μg/μL. A standard curve was made by preparing a 2-fold serial dilution series of lysate from another sample containing these spiked-in recombinant proteins using the labelling buffer of the Amersham QuickStain kit as diluent (the protein concentrations of these standards ranged between 0.0125 and 0.4 μg/μL). 10 μL of each sample (2 μg of total protein for samples treated as unknowns, and between 0.125 and 4 μg of total protein for standards) was combined with 6.25 μL of Amersham QuickStain labelling buffer, 6.25 μL of Bolt 4 × LDS Buffer, and 1 μL of undiluted Cy5 stain from the Amersham QuickStain kit. The labelling reaction was performed for 30 minutes at room temperature in the dark before it was stopped by addition of 2.5 μL of 10× Bolt Reducing Agent. All samples were then heated for 10 minutes at 70°C before they were pulse spun for 15 sec on a Heraeus Pico microfuge set to 10,000*g* and loaded onto a 12-well 1 mm thick 4-12% Bis-Tris SDS-PAGE Gel (Cat. No. NW04122, Thermo Fisher Scientific). After loading the gel, it was immediately run at 165 V and 125 mA for 50 min in 1 × Bolt MOPS SDS Running Buffer. Following SDS-PAGE, the gel was rinsed 3 × 30 sec in 1 × Bolt Transfer Buffer containing 10% methanol and then, after being wet with transfer buffer, it was fluorescently imaged using a GE Life Sciences AI600 set-up for Cy5 detection (approximately 10 minutes of image capture). The proteins were then transferred to Protran Reinforced nitrocellulose membrane (Cat. No. 10600016, GE Life Sciences) for 1 hour at 10 V, 165 mA, in 1 × Bolt Transfer Buffer containing 10% methanol. After the transfer step, the membrane was rinsed with MilliQ water, and Cy5-labelled proteins present on the membrane were fluorescently imaged using a GE AI600 (approximately 5 minutes of image capture). The membrane was rinsed in Tris-Buffered Saline with 0.1% Tween-20 (TBST) and blocked in 5% skim milk powder in TBST for 1 hour at room temperature. The membrane was then incubated with primary antibody diluted 1/1000 in 5% skim milk powder in TBST for approximately 18 hours at 4°C. The membrane was then washed 3 × 5 min in TBST and incubated with secondary anti-Mouse IgG, Horse Radish Peroxide (HRP)-conjugated antibody diluted 1/3000 in 5% skim milk powder in TBST for 1 hour at room temperature. Before detection, the membrane was washed 3 × 5 min in TBST and cut horizontally in half, and the top half containing recombinant ENPP1 was developed by immersion in Luminata Classico Western HRP substrate (Cat. No. WBLUCO100, EMD Millipore) for 1 minute at room temperature. The bottom half containing Fam3a was developed by immersion in Luminata Forte Western HRP substrate (Cat. No. WBLUF0100, EMD Millipore) for 5 minutes at room temperature. After development, excess substrate was drained off and images were immediately captured on a AI600.

Total Protein Staining and Western Blotting

Except for the Western blot shown in Figure 5, for which the method used is described above, all other membranes were dried between filter paper overnight immediately after the transfer step. Before performing Ponceau S staining, PVDF membranes were immersed in 100% methanol for 1 min followed by MilliQ water for 1 min, while nitrocellulose membranes were immersed in MilliQ water for 1 min. Membranes were then stained for 5 min in 100 mL of Ponceau S solution (0.1% w/v Ponceau S (Cat. No. P3504-50G, Sigma-Aldrich) in 5% acetic acid). Blots were rinsed 3 × 1 min in 100 mL MilliQ water and imaged using GE Life Sciences LAS-3000 or AI600 Western imaging systems. Images captured on the LAS-3000 were taken using light illumination at 1/60 sec exposure. Images on the AI600 were captured using automatic exposure on colorimetric setting. Membranes were then destained with 6 × 5 min washes in TBST or phosphate buffered saline (PBS) for chemiluminescence and infrared detection, respectively. Any blots not stained with Ponceau S were rehydrated and put through the same destaining washes as blots that were stained. Blocking of the membranes for Western blotting was performed immediately after Ponceau S destaining. Western blotting experiments were conducted at room temperature using 1 h incubations for membrane blocking, 2 h incubations for primary antibodies, and 1 h incubations for secondary antibodies. All membranes detected with infrared fluorescence were blocked in Odyssey blocking buffer (OBB) (Cat. No. 927-40000, LI-COR Bioscience), and all primary antibodies used on these membranes were diluted in OBB containing 0.1% Tween-20 to the same concentration as that used on membranes detected with chemiluminescence. In both methods, 3 × 5 min washes were conducted between primary and secondary antibody incubations as well as after secondary antibody incubation prior to imaging. Washes used TBST solution for chemiluminescence detection and PBS with 0.1% Tween-20 (PBST) for infrared detection. When using infrared detection, we performed all steps from the secondary antibody incubation in the dark. Membranes detected by chemiluminescence were developed by immersion in Luminata Classico Western HRP substrate for 1 min. Excess substrate was drained off and images were immediately captured on a LAS-3000 or AI600. Membranes for infrared detection were scanned on the LI-COR Bioscience Odyssey CLx imaging system. Both wet and dry membranes were scanned in the Odyssey CLx. Representative images of Western blot and Ponceau S staining data not shown in the main results figures are provided in supplementary materials.

Image Capture Software and Image Formats

Images captured using the LAS-3000 system were taken in Fujifilm Image Read LAS-3000 software version 2.0 and saved as raw.inf/.img files. AI600 images were taken in GE Imager Version 1.2.0 and saved as TIFF files. Westerns scanned using the Odyssey CLx were imaged using LI-COR Image Studio software version 2.1.10 and images were saved as a work area. All densitometry analyses were performed on images saved in the above formats. Images of colorimetric Ponceau S stains as well as Western blots detected with chemiluminescence, which were analysed as part of the datasets presented in Figures 1–4, were analysed using MultiGauge image analysis software version 3.0. The example Western blot data and Cy5-total protein labelling data presented in Figure 5 were analysed using Image Quant TL software version 8.1. All densitometry analyses performed on membranes scanned on the Odyssey CLx was done using Image Studio Light version 5.

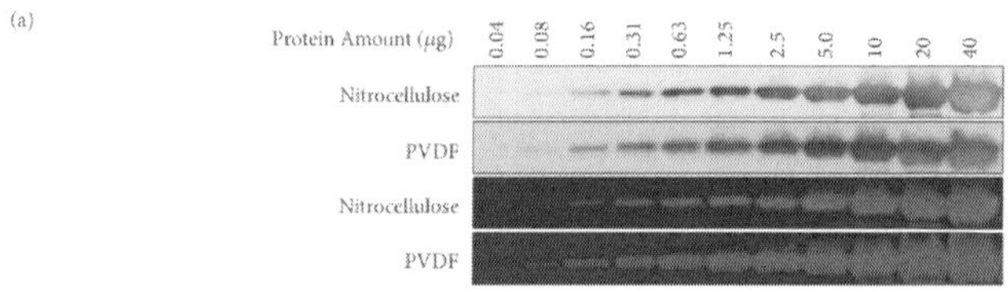

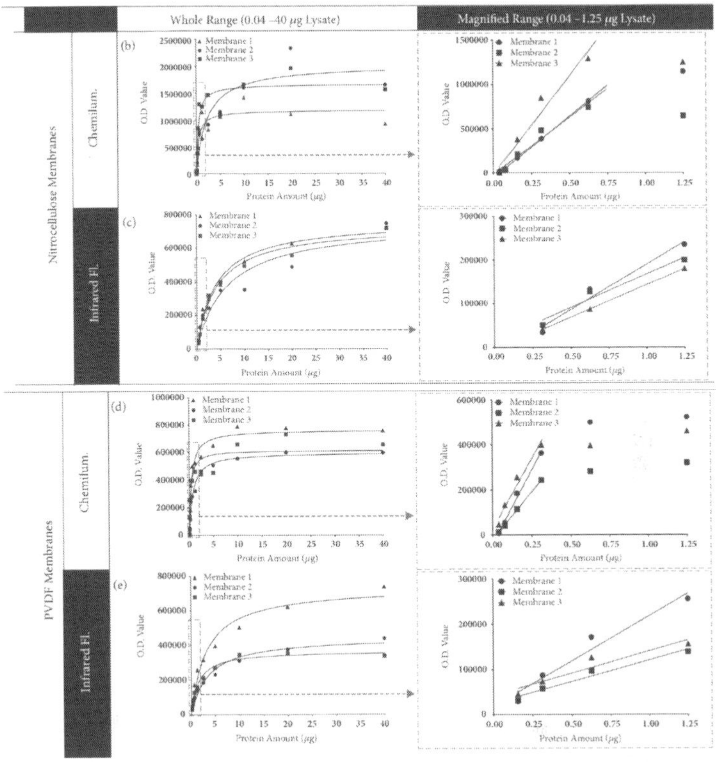

Figure 1. Detection of commonly used loading control protein SMA. Pooled pregnant human myometrial tissue homogenates extracted in 2D lysis buffer were run in 2-fold serial dilutions from 40 μg to 40 ng of total protein lysate per lane. (a) Representative Western blot images of membranes detected with chemiluminescence (black bands) or infrared fluorescence (red bands). O.D. data were calculated from band area and background-subtracted intensity for (b) nitrocellulose membranes detected by chemiluminescence, (c) nitrocellulose membranes detected by infrared fluorescence, (d) PVDF membranes detected by chemiluminescence, and (e) PVDF membranes detected by infrared fluorescence. The panels on the right side of subfigures (b-e) show the same O.D. data as each panel on the left but have been restricted to data obtained between 1.25 μg and 40 ng of total protein lysate per lane to magnify the display of these data. The area that has been magnified in the left panels are shown by red rectangles. In these magnified regions, linear regression models are fitted to the data as shown by straight lines. 3 independently prepared membranes were analysed with each detection method.

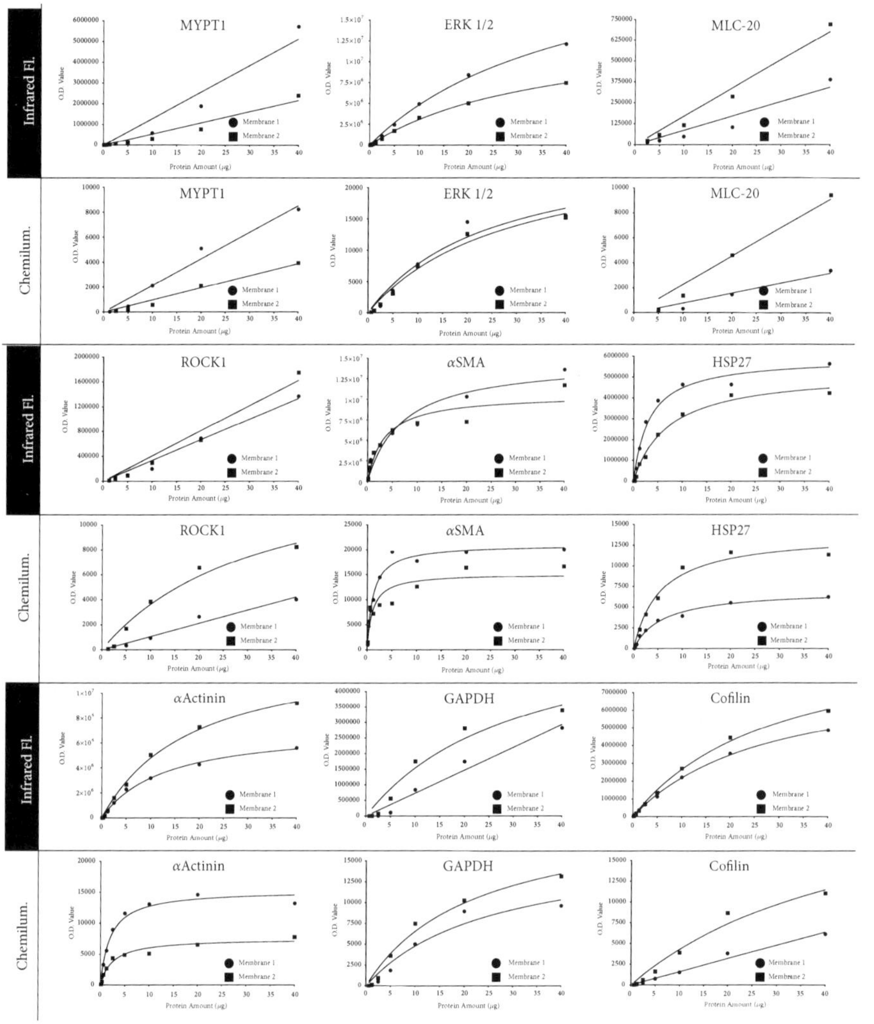

Figure 2. O.D. data linearity is protein dependent. Pooled pregnant human myometrial tissue homogenates extracted in an SDS-based lysis buffer were run in 2-fold serial dilutions from 40 μg to 160 ng of lysate. The following proteins were detected (left to right from top to bottom): MYPT1, ERK1/2, MLC-20, ROCK1, αSMA, HSP27, αActinin, GAPDH, and Cofilin. Each protein was detected with infrared fluorescence (top figure in each panel) and chemiluminescence (bottom figure in each panel). 2 independently prepared membranes were analysed with each detection method.

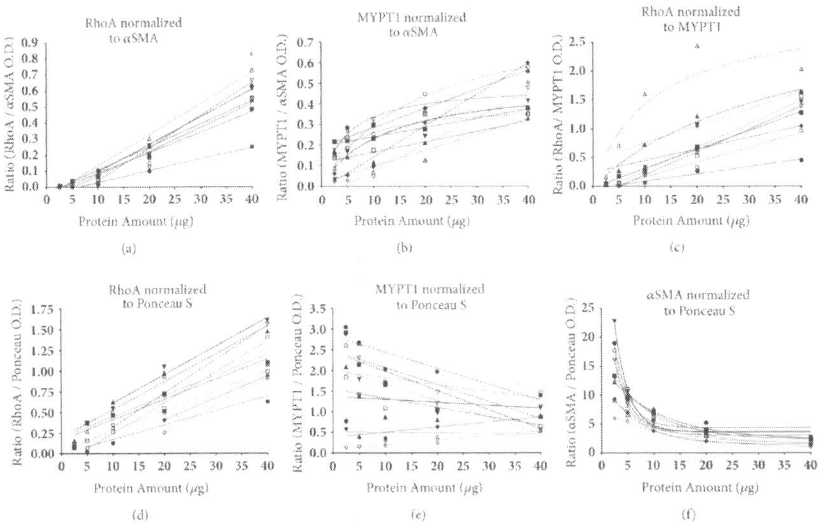

Figure 3. Assessment of errors in normalisation of Western blot densitometry data. Western blot O.D. data that were poor fits to proportional linear models were normalised to O.D. data from commonly used loading controls. Pregnant human myometrial tissue homogenates extracted in 2D lysis buffer were run in independent dilutions at 40, 20, 10, 5, and 2.5 μg of total protein lysate. A myometrial sample from a woman not-in-labour and a sample from a woman in labour were run on each membrane and this was repeated five times with samples from different women used in each repeat. Densitometry was performed by selecting the highest non-detector saturated exposure that was available. These data were normalised to the indicated loading control from the same lane in each membrane. (a) RhoA normalised to αSMA, (b) MYPT1 normalised to αSMA, (c) RhoA normalised to MYPT1, (d) RhoA normalised to Ponceau S, (e) MYPT1 normalised to Ponceau S, and (f) αSMA normalised to Ponceau S.

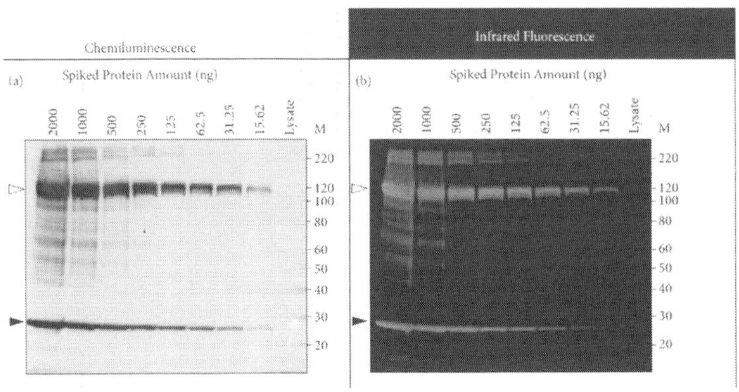

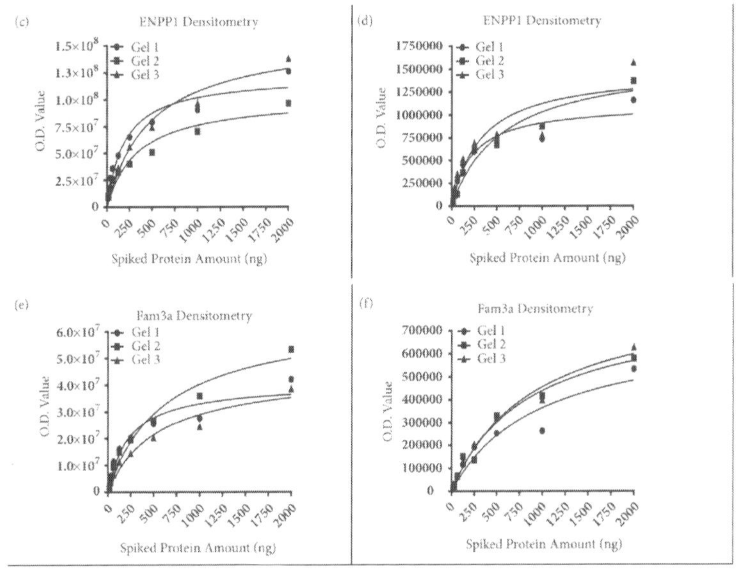

Figure 4. Densitometry analyses and representative Western blots of lysates spiked with recombinant proteins. (a) Representative blot detected using chemiluminescence. (b) Representative blot detected using infrared fluorescence. In both images the top band at approximately 120 kDa is recombinant ENPP1 (open arrow) and the bottom band at approximately 25 kDa is recombinant Fam3a (closed arrow). MagicMark XP was used for sizing (not shown). (c) ENPP1 O.D. values in membranes detected with chemiluminescence. (d) ENPP1 O.D. values in membranes detected with infrared fluorescence. (e) Fam3a O.D. values in membranes detected with chemiluminescence. (f). Fam3a O.D. values in membranes detected with infrared fluorescence. 3 independently prepared membranes were used for each detection method.

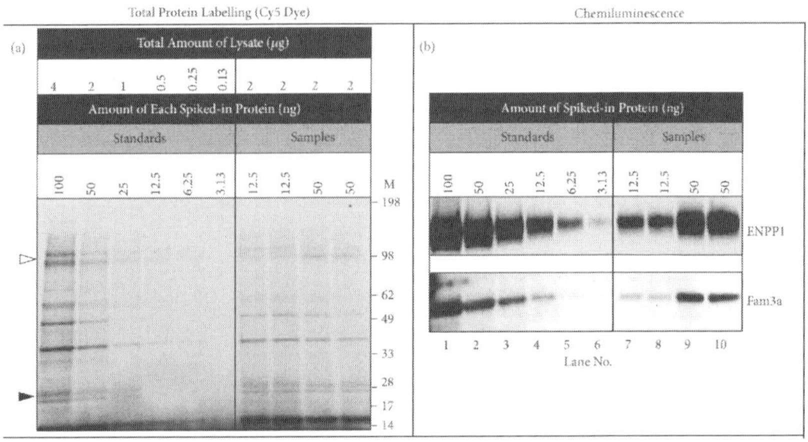

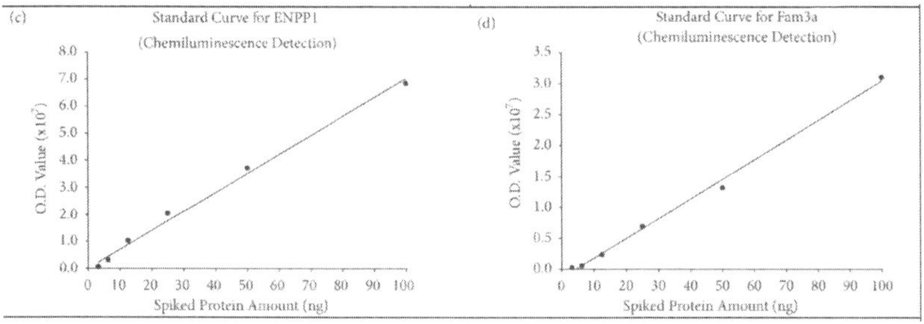

(e)

	ENPP1		Fam3a	
	Raw O.D. Data	Data Interpolated to Standard Curve (ng)	Raw O.D. Data	Data Interpolated to Standard Curve (ng)
12.5 ng Spiked-in Protein Sample (Lane 7)	6675069	9.5	2019349	10.8
12.5 ng Spiked-in Protein Sample (Lane 8)	6098493	8.7	1877634	10.3
Average of 12.5 ng Spiked-in Protein Samples	6386781	9.1	1948491	10.6
50 ng Spiked-in Protein Sample (Lane 9)	21259018	30.3	11967760	42.0
50 ng Spiked-in Protein Sample (Lane 10)	21753152	31.0	11092060	39.3
Average of 50 ng Spiked-in Protein Samples	21506085	30.6	11529910	40.6
Estimated Fold-Difference (12.5 ng vs. 50 ng Spiked-in Protein)	3.4	3.4	5.9	3.9

Figure 5. Example of a Western blotting experiment that could be used to estimate differences in protein abundance. (a) Fluorescent image of the membrane showing proteins in each lysate that were labelled with Cy5. SeeBlue Plus2 Marker was used for sizing (not shown). This image was used to estimate the total protein abundance for each sample. Arrows indicate the approximate location of ENPP1 (open arrow) and Fam3a (closed arrow). (b) Western blot images of recombinant ENPP1, and Fam3a that were used to compare their abundance. These proteins were detected by chemiluminescence. (c) The standard curve of recombinant ENPP1 O.D. values that was used to interpolate levels of recombinant ENPP1 in each sample. (d) The standard curve of recombinant Fam3a O.D. values that was used to interpolate levels of recombinant Fam3a in each sample. (e) This table shows the results of interpolating the O.D. data against the standard curve for that protein as well as the comparisons made using the raw O.D. data.

Densitometry Method

In all software packages the region of interest (ROI) encircling each band was defined manually. To ensure that the entire band was captured, the values in the lookup table were adjusted to increase contrast. This does not alter the underlying values for quantification. All bands at the correct molecular weight ± approximately 5 kDa were analysed as the signal for that target protein. In this region any overlapping visible bands in both the Western blot image and/or electrophoretogram were included to ensure that the level of background signal subtraction was appropriate to the level of background noise in the membrane and was not artificially affected by detected signal. As comparisons are made against the same sample lysate, any nonspecific overlapping bands that may be present in this region are considered to be of constant relative abundance and are unlikely to interfere with data interpretation. Any other bands visible at different molecular weights were excluded from analyses. When bands in adjacent lanes were touching due to high protein load, the boundaries of the of ROI were placed at the point of minimum thickness between the bands. The densitometry data for Ponceau S total protein stain images were obtained from all proteins visible in each entire lane. All data normalisation processes were performed by dividing the O.D. value of the target protein by the O.D. value of the chosen loading control.

Analyses Using MultiGauge Software

Images were analysed with MultiGauge software using automatic horizontal or polygonal baseline detection with settings H ratio 10% and V ratio 70%. The width of a band was defined manually. The limits of the band were chosen to be the region on each side of the curve (top/bottom of band) in which the line denoting intensity either touched or ran parallel with the line denoting baseline.

Analyses Using Image Quant TL Software

The rolling ball algorithm set to a radius of 50 pixels was used for background subtraction. The limits of each protein band were chosen to be the region on each side of the curve (top/bottom of band) in which the line denoting band intensity touched the line denoting baseline.

Analyses Using Image Studio Software

Densitometry performed in Image Studio used median local background correction with a border width of 1 unit. The location of the region for background correction was defined around the adjacent pixels of the ROI using the top/bottom only, right/left only, or the entire ROI setting. This selection was changed to prevent counting (as background) the signal from overlapping between neighbouring lanes as well as any nontarget bands 5-10 kDa away from the target protein molecular weight as these were likely to interfere with proper background subtraction if unaccounted for. Once the border region was chosen for background subtraction, this selection was kept consistent across all repeats analysing the same protein.

Statistical Analysis of Regression Fits

Regression modelling was performed using GraphPad Prism version 6. All regression models were fit using the least squares method. Comparisons between the fit of each dataset by hyperbolic and linear regression models were made using the extra-sum-of-squares F-test with an F-statistic threshold of $= 0.05$ for model selection [38].

RESULTS

O.D. Data May Be Saturated When Western Blots Are Analysed at Typically Used Amounts of Protein Lysate

2-fold serial dilutions of pooled lysates of term pregnant human myometrial tissue were separated by SDS-PAGE and transferred to nitrocellulose or PVDF membranes. These were probed with an anti-αSMA antibody and detected using chemiluminescence or infrared fluorescence (Figure 1). O.D. data were analysed for saturation by comparing a hyperbolic model against a linear model [4].

O.D. data from all membranes were better fit with hyperbolic rather than linear models, indicating that data were becoming saturated as the amount of lysate loaded into the gels increased (Figures 1(a)–1(e)). The linear region of detection occurred below approximately 0.3-1.25 μg of lysate. In the linear region of detection, O.D. data directly correspond to protein abundance only if the data fit directly proportional linear models [2, 5]. If linear O.D. data instead fit nonproportional linear models, the O.D. data do not directly correspond to protein abundance, and if this is not accounted for, any estimates of protein abundance differences between samples will be incorrect [2, 5]. In these Western blots, O.D. data above approximately 0.3-1.25 μg of lysate were nonlinear and plateaued when more than 5-20 μg of lysate was analysed. This differed depending on the type of membrane that was used. High dose hook-like effects were present in the plateaued region in all membranes detected by chemiluminescence. The nonlinear but not plateaued region of detection should only be used for quantification if O.D. data from samples are interpolated against O.D. data of a standard curve that has been produced from a calibrator sample (lysate or purified protein) run over enough dilutions to define the appropriate model [2, 4, 5, 21, 25]. This method can also be used to estimate protein abundance differences when O.D. data are linear but nonproportional [2, 5]. To control for membrane-specific effects, it is important that the dilution series of the calibrator is included on each membrane that contains samples being compared [2, 4, 5, 21, 25], and O.D. data that fall outside the range of the standard curve should not be analysed [4, 20]. Only qualitative comparisons showing the presence or absence of a protein/PTM should be made in the region of detection where O.D. data have plateaued because protein/PTM abundance differences may not be detected [3, 4, 8, 12, 13].

Visual inspection of the Western blots (Figure 1(a)) shows that band area increases with protein load. This suggested that saturation of O.D. data might be attributable to the measurement of band intensity and, therefore, that measuring band area alone may provide a solution to the saturation problem. For blots detected with chemiluminescence, we separately examined band intensity and band area. These data were both best fit with hyperbolic curves, indicating that neither measure alone surmounts the problem of O.D. data saturation.

Band Ghosting Artefacts Prevent Accurate Quantitative Comparisons

In Figure 1(a), band fading, known as "ghosting" [12, 21], can be seen above approximately 2.5 μg of lysate in nitrocellulose membranes and 10 μg of lysate in PVDF membranes detected with chemiluminescence. Ghosting was not observed in membranes detected with infrared fluorescence. This artefact results in bands appearing washed out [4, 21] and caused the aforementioned high dose hook-like effects in αSMA O.D. values that were present in Figures 1(b) and 1(d). Western blots that show ghosted bands are therefore unsuitable for quantitative comparisons because densitometry will not provide reliable measures of protein abundance [4, 21]. Since ghosting is associated with overabundance of the target protein, it should be addressed by reducing the amount of lysate that is loaded into SDS-PAGE gels [4].

Each Protein in a Sample Can Have Different Linearity

We investigated whether saturation affected the linearity of O.D. data for other myometrial proteins (Figure 2). As part of this experiment we also tested whether saturation occurred under different experimental conditions to those used for the samples shown in Figure 1. This was achieved in the analyses of the samples shown in Figure 2 by using a different human myometrial tissue lysate, a different SDS-PAGE system, supplier of nitrocellulose membrane, and chemiluminescence detection system.

Each protein analysed had different linearity and/or detectable range across the dilution series (40–0.16μg of lysate) (Figure 2). O.D. data for αACTININ, αSMA, HSP27, and ERK1/2 were affected by saturation as they were fit with hyperbolic models. MYPT1 and MLC-20 O.D. data were unaffected by saturation since they fit linear models. ROCK1, Cofilin, and GAPDH O.D. data from at least one membrane per protein were fit by hyperbolic models, and at least one membrane for each protein was fit with linear models, which indicated that the fitting regression model can be different between membranes that contain repeats of an experiment. These differences between experimental repeats may not be under control of the researcher and could be due to differences in transfer efficiency that affected the amount of proteins present on these membranes. Detection by chemiluminescence or infrared fluorescence imparted only minimal differences in the models that fit O.D. data obtained for the same protein.

Your Normalised Western Data May Be Misleading

In a different set of experiments that used dilution series of different myometrial lysates, each membrane was stained with Ponceau S to detect total protein and then probed for MYPT1, αSMA, and RhoA. Proportional linear models were not appropriate fits for most MYPT1, αSMA, and RhoA O.D. data (Supplementary Figure S7). We were therefore able to explore under real experimental conditions whether data normalisation is compromised when O.D. data deviate from proportional linear models (Figure 3).

When we normalised one of these proteins against another protein (Figures 3(a), 3(b), and 3(c)) or a protein against the Ponceau S stain (Figures 3(d), 3(e), and 3(f)), the normalised ratio was not constant, and it systematically differed across each dilution series in a manner that was specific to the proteins under investigation. These findings confirm that it is inappropriate to normalise target protein O.D. data by dividing it by the O.D. data of loading control protein(s) if the O.D. data do not fit proportional linear models [2, 7, 20, 24].

O.D. Data Nonlinearity and Normalisation Error Can Occur under Typical Western Blot Conditions

We spiked myometrial homogenates with two purified recombinant proteins, ENPP1 (~97 kDa) and Fam3a (~23 kDa). By accounting for the amount of each recombinant protein and using the unspiked lysate as a diluent, we were able to vary the concentrations of spiked-in proteins in a 2-fold serial dilution series while maintaining similar total amounts (w/v) of lysate. These samples allowed us to assess O.D. data linearity and normalisation error under conditions like how typical Western blot comparisons are performed, in which samples are compared at a single amount of lysate that is held constant between comparisons.

O.D. data from ENPP1 and Fam3a were both fit with hyperbolic models (Figure 4). There was only a small range of ENPP1 and Fam3a O.D. data that were approximately linear, which occurred below approximately 250 ng of spiked-in protein (<0.63% of spiked-in proteins in the total lysate). In each sample lane, we normalised ENPP1 to Fam3a and separately normalised ENPP1 and Fam3a to Ponceau S O.D. data (Supplementary Figure S9). When ENPP1 was normalised to Fam3a (this ratio should be constant between samples), a small normalisation error of approximately 2-fold was found across the dilution series. The Ponceau S O.D. data were similar between lanes (Supplementary Figure S8), which was expected because each sample had a similar amount of lysate loaded into the gel; hence normalisation of ENPP1 and Fam3a to Ponceau S was like dividing the raw ENPP1 and Fam3a O.D. data by a constant, and the normalised data preserved the hyperbolic pattern of the raw data. If these normalised data (ENPP1:Ponceau S or Fam3a:Ponceau S) were used to compare ENPP1 or Fam3a protein levels between these samples, the estimated differences in protein abundance would be incorrect. All normalisation errors observed in this dataset were similar in pattern and magnitude between membranes detected with chemiluminescence or infrared fluorescence.

More Accurate Estimation of Differences in Protein Abundance May Be Made after Calibration of Western Blotting Experiments

Figure 5 shows an example of how quantitation of protein abundance using Western blot O.D. data can be successfully performed. Using the data from the experiments in Figure 4 as the guide, we performed an experiment in which we compared, in duplicate, two samples that had a 4-fold difference in the abundance of spiked-in recombinant ENPP1 and Fam3a proteins. The absolute amounts of recombinant ENPP1 and Fam3a protein analysed (12.5 ng in one sample vs. 50 ng in the other) were calculated to be within the linear range of detection, and we included a standard curve covering the range of 3.25–100 ng of recombinant ENPP1 and Fam3a spiked into myometrial lysate. Comparison of the total protein abundance on the membrane (used as a loading control) suggested that it did not differ by more than approximately 0.1-fold between the samples being compared (Figure 5). This meant that it was unlikely that confounding factors affected comparisons. Thus, it would have been superfluous to put the Western blot O.D. data through a normalisation process. O.D. data for recombinant ENPP1 measured in the standards fitted a proportional linear equation (Figure 5) (equation y = 702615x, relative $R^2 = 0.993$). The standard curve is useful here as a quality control check to demonstrate that the O.D. data fit this equation [2, 3, 7]. However, in this ideal situation [2–5], the standard curve does not improve estimations of differences in protein abundance. Thus, a 3.4-fold difference in recombinant ENPP1 protein abundance was found between the two samples being compared when comparisons were made using the raw O.D. data as well as when comparisons were made after O.D. data were first-interpolated against the standard curve. In contrast, O.D. data for Fam3a measured in the standards were better fit with a nonproportional linear equation (Figure 5) (equation y = 318398x – 1411334, $R^2 = 0.996$) than a proportional linear equation ($P = 0.0433$, compared using extra-sum-of-squares F-test). A 5.9-fold estimated difference in recombinant Fam3a abundance was found between the two samples when the raw O.D. data were compared. As described above and highlighting the importance of the standard curve in these cases [2, 5], interpolation of the Fam3a O.D. data against the standard curve before comparisons were made resulted in a more accurate 3.9-fold difference in recombinant Fam3a abundance.

DISCUSSION

Our study confirms that quantification of Western blot data is not straightforward [2–4, 20]; users cannot arbitrarily select the amount of lysate to load into a gel and expect meaningful results, even when protein loading is held constant across gel lanes.

Reproducibility, detection sensitivity, data linearity, and proportionality as well as the dynamic range of detection are important parameters to assess for any Western blot method [2, 3, 20]. When we compared these parameters between detection by chemiluminescence or infrared fluorescence, we did not find differences that were large enough to recommend that one method should be used over the other. While this is in conflict with some recent studies that suggested that detection using infrared fluorescence produced superior quantitative Western blot data [4, 18], our findings confirm that it is more useful to recognise that limitations exist in the collection and analysis of all Western blot data, regardless of the detection method [3]. Thus, only with proper quality control strategies can the interpretation of quantitative Western blot data be relied upon to have biological relevance [2–5, 7, 11, 20, 21, 25].

It is critical to address detector-independent saturation that affects the linearity of Western blot O.D. data in order to ensure that meaningful results are obtained [2–21]. Multiple explanations for this type of saturation have been suggested, including that it may be due to:

- saturation of membrane binding sites [5, 11, 12], after which proteins will bind in layers that hide the detection of proteins layered underneath [11];
- limited accessibility of dyes, antibodies, and/or antigens [4, 12];
- the concentration of primary or secondary antibodies used [6, 25];
- limitations in the local concentration of HRP substrate [25];
- oxidation of HRP attached to the secondary antibodies [4, 20];
- proximity quenching of infrared fluorescence reactions [4].

Regardless of the cause of saturation, our work and the work of others [2–21, 25] indicate that it should not be assumed that O.D. data provides a direct measure of protein abundance [2, 4, 5, 7]. Linear proportionality should not be assumed even when based on similar studies, as differences in methodology that appear trivial affect the linearity of O.D. data [4, 25]. Therefore, to allow proper data interpretation and increase the confidence that findings are accurate and reproducible, the appropriate model for O.D. data should be established and presented for every protein analysed in every study [2, 3, 7, 11, 12, 20, 39]. This assessment must be performed under identical conditions to those used to analyse experimental samples, and it is also essential that the levels of all proteins being studied are similar between the lysate used for this assessment and the experimental samples [3, 7, 11, 12, 20]. This may be accomplished using dilution series of pooled lysate from all of the experimental samples [3, 7, 11, 12, 20, 39]; however, because this method averages protein levels between samples, it should be ensured that the analysis is conducted across a dilution span that is sufficient to encompass all protein amounts that may be encountered in the study [11, 39].

O.D. data are commonly nonlinear and often plateaued when high abundance proteins (often housekeeping proteins such as GAPDH, α-Tubulin, and β-Actin) were detected above 1-10 μg of lysate [2–21]. Our results, showing that αSMA and HSP27 O.D. data were nonlinear when more than approximately 1.25 μg of human myometrial tissue lysate was analysed, further question the reliability of Western blot studies that have used other high abundance proteins as loading controls. It is worth restating that it is known that GAPDH, α-Tubulin, and β-Actin are often not appropriate loading controls, as their levels change under different biological and experimental conditions at the mRNA and/or protein level [7, 18, 40–45]. These changes in abundance may be due to *in vivo* biological differences [18], may be induced by the experimental treatments under investigation [7, 41], or may be related to the *in vitro* biology of experimental models [43]. Levels of cell-type specific proteins that are used as loading controls may also change under different experimental conditions. For example, Campbell et al. [45] found that αSMA mRNA levels changed according to cell density in primary cultured rabbit aortic smooth muscle cells. Therefore, it needs to be confirmed that levels of loading control protein(s) do not change between the experimental populations being compared or, if this evidence has already been published, it should be referred to [2].

We confirmed that when O.D. data deviate from proportional linear models, the reliability of normalisation performed by dividing target protein O.D. data with loading control protein O.D. data is compromised [2, 7, 20, 24]. Our data and that of others [7, 24] show that normalisation performed under these circumstances does not account for irrelevant differences in protein/PTM abundance regardless of whether these are biologically driven or due to technical artefacts. At best it will add variance to protein/PTM abundance estimates [12] and will lead to over-or-under estimation of the magnitude of protein/PTM abundance changes [2, 4, 5, 7, 24], and at worst it can produce normalised ratios that suggest that protein/PTM abundance differs in the opposite direction to what actually occurs in the samples being analysed. Therefore, normalised data obtained under these circumstances cannot be relied upon and must be interpreted with caution as it can lead to incorrect conclusions being made [2, 20]. It is difficult to quantitatively establish when O.D. data deviate enough from proportional linear models to make this normalisation strategy unusable. Hence, we suggest that whenever this normalisation method is used (when O.D. data are approximated by proportional linear models), the investigator must demonstrate that normalisation error cannot explain their findings and the data used to reach this conclusion should be shown. In situations where nonproportional O.D. signals are detected and the data from the protein-of-interest is to be normalised to one-or-more loading control(s), the nonproportionality should be accounted for, which may be achieved using alternative normalisation strategies [24] or by interpolating data against standard curves [2, 4, 5, 21, 25].

Total protein labelling methods are commonly recommended in the literature [2, 3, 7–10, 12, 13, 15, 17, 18, 20, 22, 25, 39]. These methods include stain-free technology that labels proteins within the gel that is subsequently used for Western transfer [2, 8, 9, 39], and stains such as Ponceau S that labels proteins bound to Western membranes [8, 15, 25]. Total protein labelling can be an appropriate loading control and O.D. data obtained from total protein labelling often remain linear at lysate loads greater than 5-10 μg of protein [2, 7–9, 12, 15, 17, 18, 20, 25, 39]. It is important to consider whether total protein labelling interferes with subsequent immunodetection [9, 17]; however, this has only rarely been found [10] and is easily tested for by comparing simultaneously imaged Western blots containing identical samples, in which total protein labelling was performed for one membrane but not the other [7, 10, 12]. Unless total protein labelling has been shown to interfere with immunodetection, there are two important reasons why every Western blot membrane should be labelled for total protein and these data presented even when loading control protein(s) are also detected using antibodies:

- Most types of proteins found in cell and tissue lysates are visualised by total protein labelling methods and, therefore, these techniques allow the levels of expression of many different proteins to be estimated [9, 10, 15, 20]. In contrast, immunodetection of specific loading control proteins is currently too resource intensive to allow routine detection of more than a few proteins on each membrane; thus, immunodetection has more chance of sampling error [9, 10, 15, 20]. Reporting of total protein labelling data can often allow other researchers to best answer questions on whether different samples have similar SDS-PAGE protein separation profiles and, when appropriate, whether the amount of total protein analysed for each sample is similar [9, 10, 15, 20].

- Total protein labelling of Western membranes best shows artefacts such as degraded samples, smeared bands due to protein precipitation in the gel, and transfer errors such as air bubbles [7, 25].

It is also important to recognise that labelling all proteins in a sample does not discriminate between the sources of protein when more than one biological source is present [3]. Therefore, even if the same amount of total protein is detected between samples, it may originate from differences in cell types and/or extracellular matrix, and if these differences are not relevant to the experimental question being addressed, it can cause incorrect results [3]. While the same limitation applies for ubiquitously expressed loading control proteins detected using antibodies, this can occasionally be circumvented when examining a tissue-specific protein by using a loading control protein specifically expressed in the same tissue as the target protein [3]. To show that differences in tissue composition do not affect findings made by Western blotting when total protein labelling or ubiquitously expressed loading control proteins are used, a histological analysis should also be performed or the relevant literature should be cited [3].

CONCLUSIONS

We showed that there are many technical challenges to overcome when using Western blotting to compare the abundance of proteins found in myometrial tissue lysates. Our data support the literature and confirm that it is difficult to convert Western blotting from a qualitative technique to a quantitative technique [2–4, 7, 11, 20]. By analysing myometrial proteins including αSMA, HSP27, αActinin, and GAPDH, we confirmed that loading too much lysate into an SDS-PAGE gel can cause problems such as detector-independent O.D. data saturation, nonlinearity of O.D. data, and ghosting artefacts in subsequent Western blotting experiments [2–4, 7, 11, 20]. Compounding these problems is the limitation that the standard method of data normalisation that is used in the field only works when the O.D. data used in the normalisation calculations are directly proportional to the abundance of the corresponding proteins in the samples under study [4–6, 24]. Our data confirmed that when this is not the case, confounding differences in protein abundance arising from technical or biological origin may not be accounted for in the normalised data. In conclusion, our findings, taken together with those in the literature [2–4, 7, 11, 20], suggest that much of the Western blot data being routinely reported in the scientific community are likely to be severely compromised and are unjustly being used to support claims of biological significance. Below we have made general recommendations based on those in the literature [2–4, 7, 11, 12, 20, 25, 39]. The use of these or similar recommendations by the scientific community will lead to a much higher quality standard of Western blot data that are reported.

Recommendations

- Perform total protein labelling and image the labelled membranes. Present these images alongside all Western blot data or in supplementary material.
- Densitometry analysis should not be performed on bands showing ghosting as it is unreliable and can lead to incorrect data interpretations.
- Degraded samples showing band smearing, overloaded lanes with streaked bands due to protein precipitation in the gel, or bands with a wavy pattern indicative of overheated gels should not be used for quantitative comparisons.
- Perform Western blot assays specifically designed to establish the dynamic range of detection and determine whether O.D data fit proportionally linear, linear but nonproportional, or nonlinear models for all target protein(s)/PTM(s) being assessed, as well as all loading control(s). This critical assessment should be performed under identical technical and sample conditions to those that will ultimately be used in the study. This is ideally repeated over at least two membranes to minimise the chance that membrane-specific effects interfered with this important validation step.
- Load an experimentally justified amount of lysate. This amount should be based on calibrations performed in step (4). Quantitative comparisons should not be made in the saturated range of protein loading. If O.D. data fit linear but nonproportional or nonlinear models, this should be accounted for in all quantitative comparisons.
- Western blots on validated unchanging loading controls should be run at optimised nonsaturated conditions. This may require samples investigating loading control(s) to be run at different dilutions in different lanes and/or membranes to target proteins.

- Provide evidence that data normalisation steps are functioning for their intended purpose and are not introducing errors into the analysis.
- Use appropriate statistical analyses and statistical blocking strategies that account for inter- and intramembrane differences.

DATA AVAILABILITY

All densitometry data and representative images of Western blots and membranes labelled with total protein stains that were used to support the findings of this study are included within the supplementary materials.

CONFLICTS OF INTEREST

The authors declare that there are no conflicts of interest that may have affected this study.

ACKNOWLEDGMENTS

This research was funded by an Australian Postgraduate Award Scholarship and Research Training Scheme support provided by the University of Newcastle, Newcastle, Australia. The authors would like to thank Anne Wright for patient recruitment, the staff of the John Hunter Hospital, and the Mothers who generously consented for these studies.

REFERENCES

1. S. Ikami, T. Kobayashi, Y. Tanaka, and A. Yamaguchi, "Development of a Next-generation CCD Imager for Life Sciences Research," Fujifilm Research and Development, 2015, http://www.fujifilm.com/about/research/report/060/pdf/index/ff_rd060_003_en.pdf.
2. R. M. Murphy and G. D. Lamb, "Important considerations for protein analyses using antibody based techniques: down-sizing Western blotting up-sizes outcomes," *The Journal of Physiology*, vol. 591, no. 3, pp. 5823–5831, 2013.
3. A. A. McDonough, L. C. Veiras, J. N. Minas, and D. L. Ralph, "Considerations when quantitating protein abundance by immunoblot. American journal of physiology," *Cell Physiology*, vol. 308, no. 3, pp. C426–C433, 2015.
4. K. A. Janes, "An analysis of critical factors for quantitative immunoblotting," *Science Signaling*, vol. 8, no. 371, pp. rs2–rs2, 2015.
5. J. P. Mollica, J. S. Oakhill, G. D. Lamb, and R. M. Murphy, "Are genuine changes in protein expression being overlooked? Reassessing Western blotting," *Analytical Biochemistry*, vol. 386, no. 2, pp. 270–275, 2009.
6. A. Degasperi, M. R. Birtwistle, N. Volinsky et al., "Evaluating strategies to normalise biological replicates of Western blot data," *PLoS ONE*, vol. 9, Article ID e87293, 2014.
7. A. Hagner-McWhirter, Y. Laurin, A. Larsson, E. J. Bjerneld, and O. Ronn, "Cy5 total protein normalization in Western blot analysis," *Analytical Biochemistry*, vol. 486, pp. 54–61, 2015.
8. B. Rivero-Gutierrez, A. Anzola, O. Martinez-Augustin, and F. S. de Medina, "Stain-free detection as loading control alternative to Ponceau and housekeeping protein immunodetection in Western blotting," *Analytical Biochemistry*, vol. 467, pp. 1–3, 2014.
9. J. E. Gilda and A. V. Gomes, "Stain-Free total protein staining is a superior loading control to beta-actin for Western blots," *Analytical Biochemistry*, vol. 440, no. 2, pp. 186–188, 2013.
10. L. Zeng, J. Guo, H. Xu et al., "Direct Blue 71 staining as a destaining-free alternative loading control method for Western blotting," *Electrophoresis*, vol. 34, no. 15, pp. 2234–2239, 2013.
11. S. C. Taylor, T. Berkelman, G. Yadav, and M. Hammond, "A defined methodology for reliable quantification of Western blot data,"

Molecular Biotechnology, vol. 55, pp. 217–226, 2013.

12. G. M. Aldridge, D. M. Podrebarac, W. T. Greenough, and I. J. Weiler, "The use of total protein stains as loading controls: An alternative to high-abundance single-protein controls in semi-quantitative immunoblotting," *Journal of Neuroscience Methods*, vol. 172, no. 2, pp. 250–254, 2008.
13. J. S. Thacker, D. H. Yeung, W. R. Staines, and J. G. Mielke, "Total protein or high-abundance protein: which offers the best loading control for Western blotting?" *Analytical Biochemistry*, vol. 496, pp. 76–78, 2016.
14. A. Dittmer and J. Dittmer, "Beta-actin is not a reliable loading control in Western blot analysis," *Electrophoresis*, vol. 27, no. 14, pp. 2844-2845, 2006.
15. I. Romero-Calvo, B. Ocón, P. Martínez-Moya et al., "Reversible Ponceau staining as a loading control alternative to actin in Western blots," *Analytical Biochemistry*, vol. 401, no. 2, pp. 318–320, 2010.
16. J. R. Wisniewski and M. Mann, "A proteomics approach to the protein normalization problem: selection of unvarying proteins for MS-based proteomics and western blotting," *Journal of Proteome Research*, vol. 15, pp. 2321–2326, 2016.
17. C. Welinder and L. Ekblad, "Coomassie staining as loading control in Western blot analysis," *Journal of Proteome Research*, vol. 10, no. 3, pp. 1416–1419, 2011.
18. S. L. Eaton, S. L. Roche, M. Llavero Hurtado et al., "Total protein analysis as a reliable loading control for quantitative fluorescent western blotting," *PLoS ONE*, vol. 8, Article ID e72457, 2013.
19. J. You, C. Hodge, L. Wen, J. W. McAvoy, M. C. Madigan, and G. Sutton, "Using soybean trypsin inhibitor as an external loading control for Western blot analysis of tear proteins: Application to corneal disease," *Experimental Eye Research*, vol. 99, no. 1, pp. 55–62, 2012.
20. R. Ghosh, J. E. Gilda, and A. V. Gomes, "The necessity of and strategies for improving confidence in the accuracy of western blots," *Expert Review of Proteomics*, vol. 11, pp. 549–560, 2014.
21. O. Suzuki, M. Koura, Y. Noguchi, K. Uchio-Yamada, and J. Matsuda, "Use of sample mixtures for standard curve creation in quantitative western blots," *Journal of Experimental Animal Science*, vol. 60, no. 2, pp. 193–196, 2011.

22. A. D. Colella, N. Chegenii, M. N. Tea, I. L. Gibbins, K. A. Williams, and T. K. Chataway, "Comparison of Stain-Free gels with traditional immunoblot loading control methodology," *Analytical Biochemistry*, vol. 430, no. 2, pp. 108–110, 2012.
23. U. S. Neill, "All data are not created equal," *The Journal of Clinical Investigation*, vol. 119, no. 3, p. 424, 2009.
24. A. Pitre, Y. Pan, S. Pruett, and O. Skalli, "On the use of ratio standard curves to accurately quantitate relative changes in protein levels by Western blot," *Analytical Biochemistry*, vol. 361, no. 2, pp. 305–307, 2007.
25. F. Heidebrecht, A. Heidebrecht, I. Schulz, S.-E. Behrens, and A. Bader, "Improved semiquantitative Western blot technique with increased quantification range," *Journal of Immunological Methods*, vol. 345, no. 1-2, pp. 40–48, 2009.
26. M. Riley, P. N. Baker, R. M. Tribe, and M. J. Taggart, "Expression of scaffolding, signalling and contractile-filament proteins in human myometria: effects of pregnancy and labour," *Journal of Cellular and Molecular Medicine*, vol. 9, no. 1, pp. 122–134, 2005.
27. D. A. MacIntyre, E. K. Tyson, M. Read et al., "Contraction in human myometrium is associated with changes in small heat shock proteins," *Endocrinology*, vol. 149, no. 1, pp. 245–252, 2008.
28. J. Paul, K. Maiti, M. Read et al., "Phasic Phosphorylation of Caldesmon and ERK 1/2 during contractions in Human Myometrium," *PLoS ONE*, vol. 6, no. 6, Article ID e21542, 2011.
29. M. O'Brien, S. Carbin, J. J. Morrison, and T. J. Smith, "Decreased myometrial p160 ROCK-1 expression in obese women at term pregnancy," *Reproductive Biology and Endocrinology*, vol. 11, no. 1, 2013.
30. H. N. Aguilar, B. Zielnik, C. N. Tracey, and B. F. Mitchell, "Quantification of rapid myosin regulatory light chain phosphorylation using high-throughput in-cell western assays: comparison to western immunoblots," *PloS One*, vol. 5, no. 4, 2010.
31. J. Lartey, M. Smith, J. Pawade, B. Strachan, H. Mellor, and A. López Bernal, "Up-regulation of myometrial RHO effector proteins (PKN1 and DIAPH1) and CPI-17 (PPP1R14A) phosphorylation in human pregnancy is associated with increased GTP-RHOA in spontaneous preterm labor," *Biology of Reproduction*, vol. 76, no. 6, pp. 971–982, 2007.

32. C. A. Hudson, K. J. Heesom, and A. L. Bernal, "Phasic contractions of isolated human myometrium are associated with rho-kinase (ROCK)-dependent phosphorylation of myosin phosphatase-targeting subunit (MYPT1)," *Molecular Human Reproduction*, vol. 18, no. 5, pp. 265–279, 2012.
33. Y. Li, M. Reznichenko, R. M. Tribe et al., "Stretch activates human myometrium via ERK, caldesmon and focal adhesion signaling," *PLoS ONE*, vol. 4, Article ID e7489, 2009.
34. M. Baker, "Reproducibility crisis: blame it on the antibodies," *Nature*, vol. 521, pp. 274–276, 2015.
35. A. Bradbury and A. Pluckthun, "Reproducibility: standardize antibodies used in research," *Nature*, vol. 518, pp. 27–29, 2015.
36. A. Schonbrunn, "Editorial: Antibody can get it right: confronting problems of antibody specificity and irreproducibility," *Molecular endocrinology (Baltimore, Md.)*, vol. 28, no. 9, pp. 1403–1407, 2014.
37. K. M. Yamada and A. Hall, "Reproducibility and cell biology," *The Journal of Cell Biology*, vol. 209, pp. 191–193, 2015.
38. H. J. Motulsky, "How The F-Test Works to Compare Models," GraphPad Curve Fitting Guide, 2018, https://www.graphpad.com/guides/prism/7/curve-fitting/index.htm?reg_howtheftestworks.htm.
39. S. C. Taylor and A. Posch, "The design of a quantitative western blot experiment," *BioMed Research International*, Article ID 361590, 2014.
40. B. Knoops and J.-N. Octave, "α1-tubulin mRNA level is increased during neurite outgrowth of NG 108-15 cells but not during neurite outgrowth inhibition by CNS myelin," *NeuroReport*, vol. 8, no. 3, pp. 795–798, 1997.
41. L. Li, Y. Yan, H. Xu, T. Qu, and B. Wang, "Selection of reference genes for gene expression studies in ultraviolet B-irradiated human skin fibroblasts using quantitative real-time PCR," *BMC Molecular Biology*, vol. 12, no. 8, 2011.
42. M. Rocha-Martins, B. Njaine, and M. S. Silveira, "Avoiding pitfalls of internal controls: validation of reference genes for analysis by qRT-PCR and Western blot throughout rat retinal development," *PloS One*, vol. 7, Article ID e43028, 2012.
43. S. Greer, R. Honeywell, M. Geletu, R. Arulanandam, and L. Raptis, "Housekeeping genes; expression levels may change with density of

cultured cells," *Journal of Immunological Methods*, vol. 355, no. 1-2, pp. 76–79, 2010.

44. D. Lanoix, J. St-Pierre, A.-A. Lacasse, M. Viau, J. Lafond, and C. Vaillancourt, "Stability of reference proteins in human placenta: general protein stains are the benchmark," *Placenta*, vol. 33, no. 3, pp. 151–156, 2012.
45. J. H. Campbell, O. Kocher, O. Skalli, G. Gabbiani, and G. R. Campbell, "Cytodifferentiation and expression of alpha-smooth muscle actin mRNA and protein during primary culture of aortic smooth muscle cells, Correlation with cell density and proliferative state," *Arteriosclerosis*, vol. 9, no. 5, pp. 633–643, 1989.

6

Application of Monoclonal Antibodies against Bioactive Natural Products: Eastern Blotting and Preparation of Knockout Extract

Hiroyuki Tanaka,[1] Osamu Morinaga,[2] Takuhiro Uto,[2] Shunsuke Fuji,[3] Frederick Asare Aboagye,[4] Nguyen Huu Tung,[2] Xiao Wei Li,[5] Waraporn Putalun,[6] and Yukihiro Shoyama[2]

[1]Graduate School of Pharmaceutical Science, Kyushu University, 3-1-1 Maidashi, Higashi-ku, Fukuoka 812-0855, Japan

[2] Faculty of Pharmaceutical Science, Nagasaki International University, 2827-7 Huis Ten Bosch, Sasebo 859-3298, Japan

[3] Faculty of Health Management, Nagasaki International University, 2827-7 Huis Ten Bosch, Sasebo 859-3298, Japan

[4]Phytochemistry Department, Centre for Scientific Research into Plant Medicine, University of Ghana, MampongAkuapem, Ghana

[5] State Key Laboratory of Natural and Biomimetic Drugs, School of Pharmaceutical Sciences, Peking University, No.38 Xue-yuan Road, Haidian District, Beijing 100191, China

[6] Faculty of Pharmaceutical Science, Khon Kaen University, Khon Kaen 40002, Thailand

Citation: Application of Monoclonal Antibodies against Bioactive Natural Products: Eastern Blotting and Preparation of Knockout Extract. Hiroyuki Tanaka, Osamu Morinaga, Takuhiro Uto, Shunsuke Fuji, Frederick Asare Aboagye, Nguyen Huu Tung, Xiao Wei Li, Waraporn Putalun, and Yukihiro Shoyama. Hindawi Publishing Corporation International Journal of Analytical Chemistry Volume 2012, Article ID 260425, 9 pages doi:10.1155/2012/260425

ABSTRACT

Matrix-assisted laser desorption/ionization (MALDI) tof mass spectrometry was used for the confirmation of hapten number in synthesized antigen. As application of MAb, the MAbs against ginsenosides and glycyrrhizin have been prepared resulting in the development of two new techniques that we named the eastern blotting method and the knockout extract preparation. In eastern blotting technique, glycosides like ginsenosides and glycyrrhizin separated by silica gel TLC were blotted to PVDF membrane that was treated with a $NaIO_4$ solution followed by BSA resulted in glycoside-BSA conjugate on a PVDF membrane. The blotted spots were stained by MAb. Double staining of eastern blotting for ginsenosides using antiginsenoside Rb_1 and Rg_1 MAbs promoted complete identification of ginsenosides in *Panax* species. The immunoaffinity concentration of glycyrrhizin was determined by immunoaffinity column conjugated with antiglycyrrhizin MAb resulting in the glycyrrhizin-knockout extract, which was determined by the synergic effect with glycyrrhizin on NO production using the cell line.

INTRODUCTION

Immunoassay systems using monoclonal antibody (MAb) against drugs and small molecular weight bioactive compounds have become an important tool for studies on receptor binding analysis, enzyme assay, and quantitative and/or qualitative analytical techniques in animals or plants. The immunoblotting method is based on the western blotting technique that utilizes the antigen-antibody binding properties and provides a specific and sensitive detection of higher molecule analytes like peptides and proteins. In our ongoing study on MAb, previously we prepared various kinds of MAb against natural products like forskolin [1], solamargine [2], crocin [3], marihuana compounds [4], opium alkaloids [5], ginsenosides [6, 7], berberine [8], sennosides [9], paeoniflorin [10], glycyrrhizin [11, 12], ginkgolic acid [13], aconitine alkaloid [14], baicalin [15], and so on and developed individual competitive enzyme-linked immunosorbent assay (ELISA) as a highly sensitive, specific, and simple methodology.

The confirmation of hapten number in synthesized antigens is most important in the first stage of MAb preparation. Therefore, its determination method will be discussed first of all. As an application of MAb, the MAb against ginsenosides and glycyrrhizin has been prepared resulting in the development of two new techniques that we have named the eastern blotting method [12] and the knockout extract preparation [16]. They will be introduced in this paper.

PREPARATION OF MAB AGAINST NATURAL PRODUCTS

Various methods have been employed for the determination of natural products. They include spectral methods such as infrared (IR), nuclear magnetic resonance (NMR), and circular dichroism (CD) and other chromatographic methods such as ion chromatography (IC), capillary electrophoresis (CE), and high-speed counter current chromatography (HSCCC), and so on. Compared to TLC, GLC and HPLC methods, the ELISA method was more sensitive and selective. Moreover, it is possible to study a large number of natural products. Since natural product extracts consist of various chemical constituents (e.g., licorice contains 470 components or more), in general, some pretreatment is necessary for HPLC and other chromatographic analysis methods. ELISA, however, can determine the concentration of components directly without any pretreatment. Therefore, ELISA was used to measure the concentration of ginsenoside Rb_1 in ginseng and traditional Chinese medicines (TCMs).

Analytical Method for Determination of Hapten Number in Antigen, Hapten-Carrier Protein Conjugate

For production of MAb, synthesis of hapten, which is derived from an immune antigen and linker bridge, and the carrier protein conjugate is necessary. There had been no direct and appropriate methods for the determination of haptens conjugated with carrier proteins without differential UV analysis, radiochemical, or chemical methods. Therefore, immunization by the injection of hapten-carrier protein conjugate was unreliable. Wengatz et al. [17] determined the hapten density of immunoconjugates by matrix-assisted UV laser desorption/ionization mass spectrometry. We also reported the direct analytical method of hapten and carrier protein conjugates by a matrix-assisted laser desorption/ionization mass spectrometry (MALDI) tof mass spectrometry using an internal standard [18–20].

Figure 1 shows the MALDI tof mass spectrum of the most pharmacologically active marihuana compound, tetrahydrocannabinolic acid (THCA)-bovine serum albumin (BSA) conjugate, and BSA used as an internal standard [20]. This shows only the singly, doubly, and triply ionized molecule ions of the intact conjugate. The sharp peak at m/z 66,465 is the $[M + H]^+$ peak of BSA. A small $[M + H]^+$ peak of the THCA-BSA conjugate is at m/z 70,792, indicating that the calculated molecular mass of the THCA-BSA conjugate is 70,581 using a calculated molecular mass of 66,267 for BSA. The calculated molecule mass of the THCA moiety is 4,314. From this result, 12.7 molecules of THCA are combined with BSA [20]. Since this method is suitable for small molecule natural products, we had been analyzing the hapten number of all natural products for MAbs including glycosides like ginsenosides and glycyrrhizin.

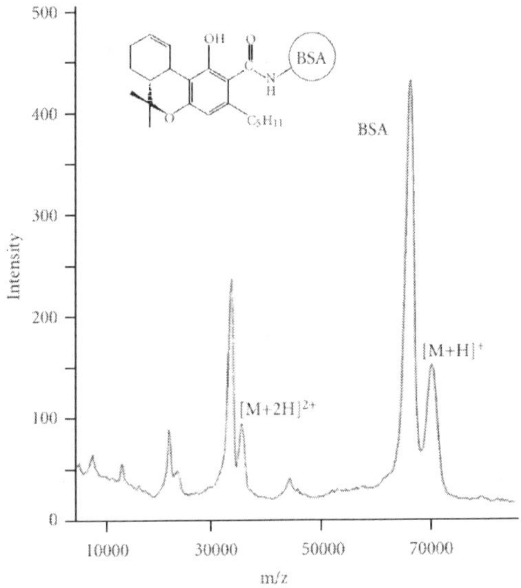

Figure 1. Matrix-assisted laser desorption/ionization tof mass spectrometry of tetrahydrocannabinolic-acid-BSA conjugate. [M + H] indicates the molecular weight of the conjugate, from which the hapten number can be calculated.

Preparation of MAb against Ginsenosides and ELISA as an Assay System

Ginseng, the crude drug of *Panax ginseng*, is one of the most important natural medicines in many countries. It has been used to enhance stamina and capacity to cope with fatigue and physical stress and as a tonic against cancers, disturbances of the central nervous system (memory, learning, and behavior), hypothermia, carbohydrate and lipid metabolism, immune function, the cardiovascular system, and radioprotection [21]. It contains more than 30 kinds of dammarane and oleanane saponins considered to be pharmacologically active components. Ginsenoside Rb_1 is the main saponin in ginseng. However, since the concentration in the ginseng root or the root extract varies depending on the method of extraction, subsequent treatment, or even the season of its collection [22], standardization of quality is required. For this purpose, we have prepared anti-ginsenoside Rb_1 [6] and Rg_1 MAbs [7]. The immunoassay system using MAb is not frequently used for naturally occurring smaller molecular weight bioactive compounds. Preparation of MAbs is difficult, but is one of the most important steps for the analysis of natural products. As a typical natural product, the preparation of MAb against the ginseng saponin ginsenoside Rb_1 will be discussed.

A hybridoma-producing MAb reactive to ginsenoside Rb_1 was obtained by the general procedure and classified into IgG2b which had light chains. The reactivity of IgG type MAb, 9G7 was tested by varying antibody concentration and by performing a dilution curve. The antibody concentration was selected for competitive ELISA. The free MAb following competition is bound to polystyrene microtiter plates precoated with ginsenoside Rb_1-HSA. Under these conditions, the full measurement range of the assay extends from 20 to 400 ng/mL [6].

Cross-reactivity is the most important factor in determining the value of an antibody. Since the ELISA for ginsenoside Rb_1 was established for phytochemical investigations involving crude plant extracts, the assay specificity was checked by determining the cross-reactivity of the MAb with various related compounds. The cross-reactivity data of MAb that was obtained were examined by competitive ELISA and calculated using picomole amounts of ginsenoside Rb_1. The cross-reactivity of ginsenoside Rc and Rd, which possess a diglucose moiety attached to the C-3 hydroxy group, was weak compared to ginsenoside Rb_1 (0.024 and 0.020%, resp.). Ginsenoside Re and Rg_1 showed no cross-reactivity (less than 0.005%). It is evident that the MAb reacted only with a small number of structurally related ginsenoside Rb_1 molecules, and very weakly, and did not react with other steroidal compounds like glycyrrhizin, digitoxin, tigogenin, tigonin, and solamargine.

In our ongoing studies on MAbs against ginseng saponins, anti-ginsenoside Rg_1 MAb [7] and Re [23] have been prepared and their ELISA was set up. Anti-ginsenoside Rg_1 MAb was also highly specific like anti-ginsenoside Rb_1. On the other hand, anti-ginsenoside Re MAb showed wide cross-reactivity. Therefore, the MAb can be used for the analysis for the total ginsenoside concentration.

Application of MAb in the Natural Products Field

Although western blotting is a common assay methodology for high-molecular-weight-substances, this method has not been employed for small molecules, as direct immunostaining of such compounds on a TLC plate is as yet unknown. Therefore, a new method for such small-molecular compounds is required. Moreover, if small molecules can be blotted to a membrane, fixing them also requires a new methodology. Previously, we succeeded in separating small-molecule compounds such as solasodine glycosides into a part of an epitope and fixing on the membrane [24], as follows.

New Staining Method for Ginsenosides, "Eastern Blotting"

Figure 2 shows the H_2SO_4 staining and eastern blotting of ginsenoside standards and TCM using anti-ginsenoside Rb_1 MAb. It is impossible to determine the ginsenosides by TLC staining by H_2SO_4 as indicated in Figure 2(a). On the other hand, clear staining of ginsenoside Rb_1 occurred by eastern blotting, as indicated in Figure 2(b). Furthermore, it became evident that Jigengtang and Dahuanggancaotang prescriptions that did not contain ginseng, as indicated by the absence of a ginsenoside Rb_1 band. The eastern blotting method was considerably more sensitive than that of H_2SO_4 staining. The H_2SO_4 staining detected all standard compounds. The eastern blotting indicated only limited staining of ginsenoside Rb_1, Rc, and Rd, whose cross-reactivities were under 0.02% as shown in Figure 2(b). We suggest that an aglycone, protopanaxadiol, and a part of the sugars may be of importance to the immunization and may function as an epitope for the structure of ginsenosides. In addition, it is suggested that the specific reactivity of sugar moieties in the ginsenoside molecule against anti-ginsenoside Rb_1 MAb may be modified by the $NaIO_4$ treatment of ginsenosides on the PVDF membrane, causing ginsenoside Rc and Rd to become detectable by eastern blotting.

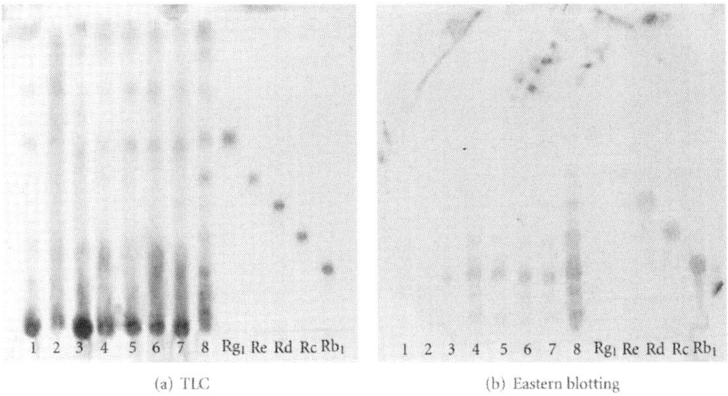

Figure 2. Eastern blotting of ginsenosides in traditional Chinese medicine formulas using anti-ginsenoside Rb1 MAb. Samples (1) Jigengtang, (2) Dahuanggancaotang, (3) Nenshenyangyongtang, (4) Sijunzitang, (5) Nenshentang, (6) Buanxiaxiexintang, (7) Xiaocaihutang, and (8) Crude ginseng extract. Lanes 1 and 2 do not contain ginseng. Standard of ginsenosides indicated, ginsenosidr-Rg_1, -Re, -Rd, -Rc, and -Rb_1, respectively, from the upper spot.

Application of the eastern blotting method for the detection of glycyrrhizin in the serum samples was investigated. In general, it is difficult to detect glycyrrhizin in serum due to the large amount of impurities. The developed bands of impurities lapped over the band of glycyrrhizin, and we failed to find a proper developing solvent system that could separate glycyrrhizin and the impurities apart from each other clearly on the TLC plate. Figure 3 shows the detection of glycyrrhizin by the eastern blotting technique in the rat serum samples. Glycyrrhizin could not be identified on the TLC plate stained by H_2SO_4 through many bands that were detected (Figure 3(a)). On the other hand, eastern blotting clearly shows the band of glycyrrhizin even after 1 h (Figure 3(b)). Although the sensitivity of the eastern blotting method was greatly affected by the impurities, the detection limit was still at the nanogram level. The results proved that the eastern blotting technique could be a unique method for identifying glycyrrhizin against a background of a large amount of impurities.

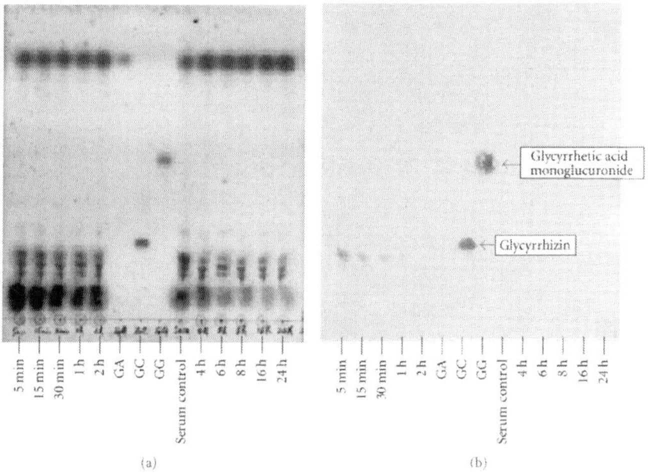

Figure 3. Eastern blotting profile of glycyrrhizin in rat serum after injection of glycyrrhizin (a) TLC of rat-serum-injected glycyrrhizin TLC was developed with *n*-BuOH-H_2O-AcOH (7 : 2 : 1). Spots were detected by 10% H_2SO_4. (b) Eastern blotting of rat-serum-injected glycyrrhizin; the band of glycyrrhizin was detected until 1 h after injection. GA: glycyhrrhetinic acid, GC: glycyrrhizin, and GG: glycyrrhetinic acid monoglucuronide.

When the mixture of anti-ginsenoside Rb_1 and Rg_1 MAbs and the pair of substrates were tested for staining for ginsenosides, all ginsenosides, ginsenoside Rb_1, -Rc, -Rd, -Re, and -Rg_1 were stained blue although the purple color staining for ginsenoside Rg_1 was expected because 3-amino-9-ethylcarbazole and 4-chloro-1-naphotol might be different. Therefore, we performed successive staining of the membrane using anti-ginsenoside Rg_1 and then anti-ginsenoside Rb_1. Finally, we performed the double staining of ginsenosides indicating that ginsenoside Rg_1 and ginsenoside Re were stained purple and the other blue, as indicated in Figure 4. From this result, both antibodies can distinguish the individual aglycones, protopanaxatriol, and protopanaxadiol. For this application, the crude extract of various *Panax* species was analyzed by the newly developed double staining system. Major ginsenosides can be determined clearly by the double staining method, as indicated in Figure 4.

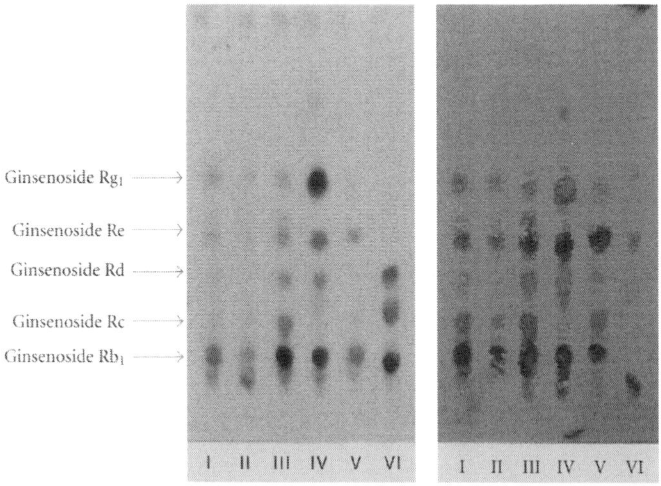

(a) (b)

Figure 4. Double Eastern blotting staining of ginsenosides contained in various ginseng samples using anti-ginseenoside-Rb1 and anti-ginsenoside-Rg_1 MAbs. (a) TLC profile stained by sulfuric acid. (b) Eastern blotting by anti-ginsenoside-Rb_1 and anti-ginseside-Rg_1 monoclonal antibodies I, II, III, IV, V, and VI indicated white ginseng, red ginseng, fibrous ginseng (*P. ginseng*), *P. notoginseng*, *P. quinquefolius* and *P. japonicus*, respectively. Upper purple color spots and lower blue color spots were stained by anti-ginsenoside-Rg_1, and anti-ginsenoside-Rb_1 monoclonal antibodies, respectively.

Therefore, it is suggested that the staining color shows the pharmacological activity. As shown in Figure 4(b), the purple bands indicate ginsenosides which have protopanaxatriol as an aglycone and stimulation activity for the central nervous system (CNS). On the other hand, the blue color indicates ginsenosides containing protopanaxadiol as an aglycone that possess a depression effect on the CNS. Moreover, the Rf value of ginsenosides roughly suggests the number of sugars attached to the aglycone. Both analyses make it possible to jointly identify which aglycone attaches and how many sugars it possesses, leading to the structure of the ginsenosides. In fact, three kinds of ginsenosides possessing protopanaxadiol-ginsenoside Rh_1, -Rf, and 20-*O*-glucoginsenoside Rf in *P. ginseng* root were determined by coloring and Rf value by comparing them with the structures reported in the previous paper [25].

Figure 5(a) indicates immunolocalization of glycyrrhizin in a licorice root slice using anti-glycyrrhizin MAb as another application of the eastern blotting method. The phloem (↔) contained a higher concentration of glycyrrhizin than the xylem and cork part [26]. On the other hand, no staining occurred on ginseng (Figure 5(c)).

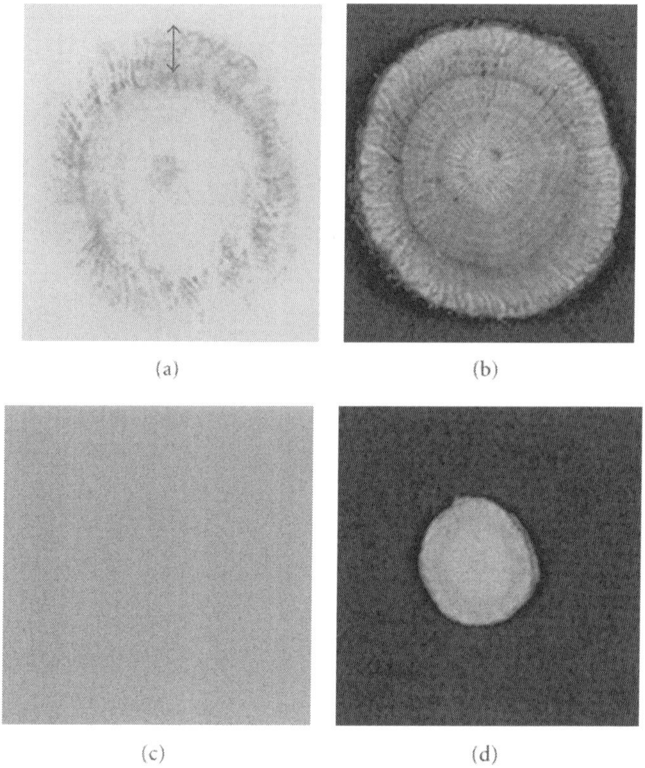

Figure 5. Immunocytolocalization of glycyrrhizin in fresh licorice root using anti-glycyrrhizin MAb. (a) Eastern blotting of fresh licorice slice, (b) fresh licorice slice, (c) Eastern blotting of fresh ginseng slice, and (d) fresh ginseng slice.

In the earlier experiments, we carried out the blotted staining on PVDF membrane using MAb on solasodine glycosides and called it western blotting [24]. Now we have applied this new methodology to licorice glycoside, glycyrrhizin and named it eastern blotting [12] for studying ginsenosides [23], saikosaponin [27], and so on.

Immunoaffinity Concentration and One-Step Purification of Ginsenoside Rb$_1$ by Immunoaffinity Column

A crude extract of *P. ginseng* roots was loaded onto the immunoaffinity column and washed with the washing solution of phosphate buffer. Figure

6 shows the fraction 1–8 containing overloaded ginsenoside Rb_1. The other ginsenosides Rg_1, Rc, Re. and Rd were also detected in these fractions by eastern blotting (data not shown). A sharp peak appeared around fraction 20–24 eluted by acetate buffer containing KSCN and methanol to give pure ginsenoside Rb_1.

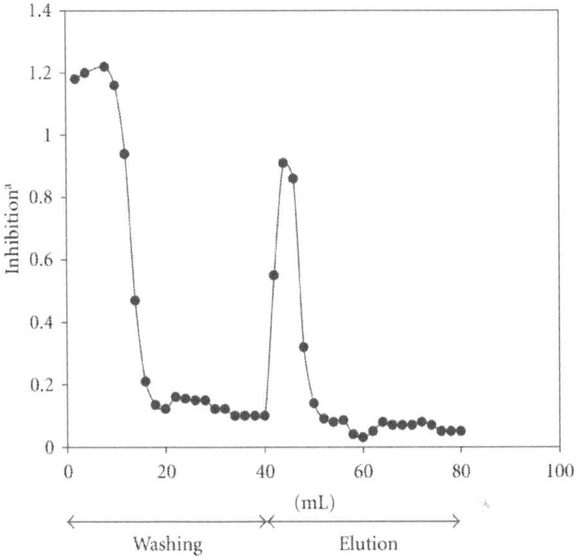

Figure 6. Elution profile of *Panax ginseng* crude extract on immunoaffinity column monitoring by ELISA using anti-ginsenoside-Rb_1 MAb. The column was washed by phosphate buffer and then eluted by HOAc buffer containing KSCN and MeOH. Individual fractions (2 mL) were assayed by ELISA using anti-ginsenoside-Rb_1 MAb.

Overloaded ginsenoside Rb_1 was repeatedly immunoaffinity column choromatographed to separate ginsenoside Rb_1 completely [28]. The antibody was stable when exposed to the eluent, and the immunoaffinity column showed almost no decrease in capacity after repeated use more than 10 times under the same conditions, as was reported for a one-step separation of forskolin from a crude extract of *Coleus forskohlii* root [29]. This methodology is effective for the rapid and simple purification of ginsenoside Rb_1 and may open up a wide field of comparable studies with other families of saponins for which an acceptable method for one-step separation has not yet been developed. Furthermore, to separate the total ginseng saponins, a wide cross-reactive MAb against ginsenoside like anti-ginsenoside Re MAb could be designed, as was done for the total solasodine glycosides by an immunoaffinity column using an anti-solamargine MAb [30]. A combination of immunoaffinity column chromatography, Eastern blotting and ELISA could be used to survey low concentrations of ginsenoside Rb_1 of plant origin and/or in experimental animals and humans. In fact, we have succeeded in the isolation of ginsenoside Rb_1 from a different plant, *Kalopanax pictus* Nakai, which was not known previously to contain ginsenosides, using this combination of methods [31].

Newly Established Knockout Extract for Glycyrrhizin

Glycyrrhizin has been reported to possess numerous pharmacological effects like anti-inflammation [32], antiulcer [33], antitumor [34], antiallergy [35], and hepatoprotective activities [36]. Although glycyrrhizin is supposed to be a major active principle in licorice crude extract, a large number of studies have demonstrated that the licorice extract is rich in bioactive compounds other than glycyrrhizin such as triterpenes, flavonoids, and their aglycones [37]. In order to confirm the effect of glycyrrhizin in TCM, previously we purified glycyrrhizin from TCM using an immunoaffinity column conjugated with anti-glycyrrhizin MAb [38]. In this section, one-step purification of glycyrrhizin and its function in the licorice crude extract have been indicated as previously discussed about ginsenoside Rb_1.

To eliminate glycyrrhizin from licorice extract, 12 mg of licorice extract (glycyrrhizin content: 1275.0 μg) in loading buffer was applied on the anti-glycyrrhizin MAb immunoaffinity column, and then the loading buffer was continuously circulated through the column to enhance the binding efficiency. After overnight circulation at 4°C, the unbound column fraction was collected. The column was washed, and then bound column fraction was eluted by elution solvent. Glycyrrhizin concentration of bound column fraction was 1269.26 μg of glycyrrhizin (99.5% of the loading glycyrrhizin). On the other hand, glycyrrhizin content of unbound fraction was 3.50 μg (0.27% of the loading glycyrrhizin). These data indicated that anti-glycyrrhizin MAb immunoaffinity column could eliminate glycyrrhizin from licorice crude extract with high efficiency. From this evidence, we named this fraction knockout extract [16, 39].

To further characterize glycyrrhizin knockout extract, the TLC analysis and eastern blotting were performed. As shown in Figure 7(a), several spots including glycyrrhizin were detected in licorice extract. However, the spots of glycyrrhizin were undetected in glycyrrhizin-knockout extract, although all other spots were clearly detected (Figure 7(a), lane 2). eastern blotting by anti-glycyrrhizin MAb demonstrated that glycyrrhizin was detected in licorice extract, but the spot of glycyrrhizin was invisible in unbound fraction (Figure 7(b), lane 2). Therefore, these data suggest that glycyrrhizin was specifically eliminated from licorice extract by anti-glycyrrhizin MAb binding immunoaffinity column.

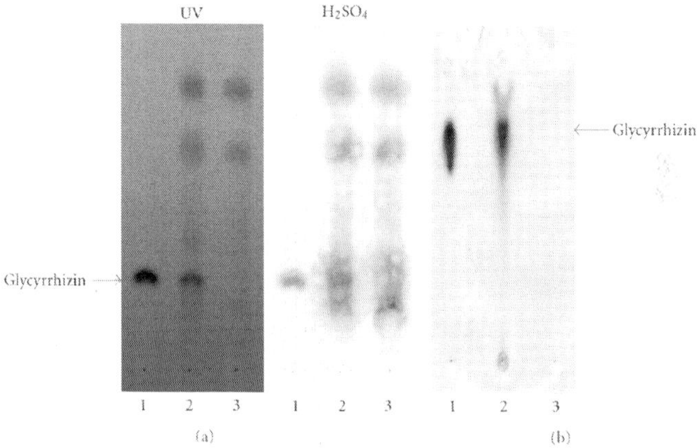

Figure 7. Preparation of knock-out extract eliminated glycyrrhizin from licorice crude extract using immunoaffinity column conjugated with anti-glycyrrhizin MAb. Lines 1, 2, and 3 indicate glycyrrhizin, crude extract, and knock-out extract, respectively.

This technique is useful approach to clarify the multicomponent interaction with the specific protein such as enzymes, but it is unsuitable to investigate the interaction between one principal component and other components in herbal medicines. To investigate the role of one principal compound in the crude extract, we prepared knockout extract by removing target compound from the crude extract using high-specific MAb immunoaffinity column. By using this approach, it may become possible to determine the potential function of one natural compound on the crude extract or TCM by *in vitro* and *in vivo* assays.

Nitric oxide (NO) is free radical with multiple physiological functions, such as vasodilatation, neurotransmission, and inflammation [40]. During inflammatory process, large amount of NO is produced by inducible nitric oxide synthase (iNOS) by inflammatory cytokines and/or bacterial lipopolysaccharide (LPS) in various cell types including macrophages [41]. Overproduction of NO by iNOS triggers the pathogenesis of septic shock and organ destruction in certain inflammatory and autoimmune diseases [42–44]. Therefore, the inhibition of NO production by blocking iNOS expression may be useful strategy for the treatment of various inflammatory diseases.

To determine whether licorice crude extract inhibits NO production in LPS-stimulated mouse RAW264 macrophages, cells were treated with various concentrations of licorice extract for 30 min and then stimulated with LPS. LPS caused a dramatic increase in NO production, and this induction was inhibited in a dose-dependent manner by treatment with licorice extract.

We next investigated the inhibitory effect of glycyrrhizin-knockout extract and the combined treatment with glycyrrhizin-knockout extract and glycyrrhizin on NO production. Since 100 μg/mL of licorice extract contains 10.6 μg of glycyrrhizin, the cells were pretreated with licorice extract (100 μg/mL), glycyrrhizin-knockout extract (89.4 μg/mL), or combination of glycyrrhizin-knockout extract (89.4 μg/mL) and glycyrrhizin (10.6 μg/mL). Figure 8(a) indicated that treatment of licorice extract leads to a marked suppression of NO production as compared to LPS treatment [inhibition ratio (IR) = 57.7%]. Interestingly, although glycyrrhizin alone could not block NO production, the inhibitory effect of glycyrrhizin-knockout extract was decreased compared with licorice extract (IR = 17.8%). Moreover, the combined treatment with glycyrrhizin-knockout extract and glycyrrhizin significantly improved the inhibition of NO production (IR = 33.5%). To determine whether the combinational effect of glycyrrhizin-knockout extract and glycyrrhizin was related to iNOS expression, we performed western blotting (Figure 8(b)). Although the licorice extract strongly suppressed LPS-induced iNOS expression, treatment of glycyrrhizin-knockout extract reduced this effect [45].

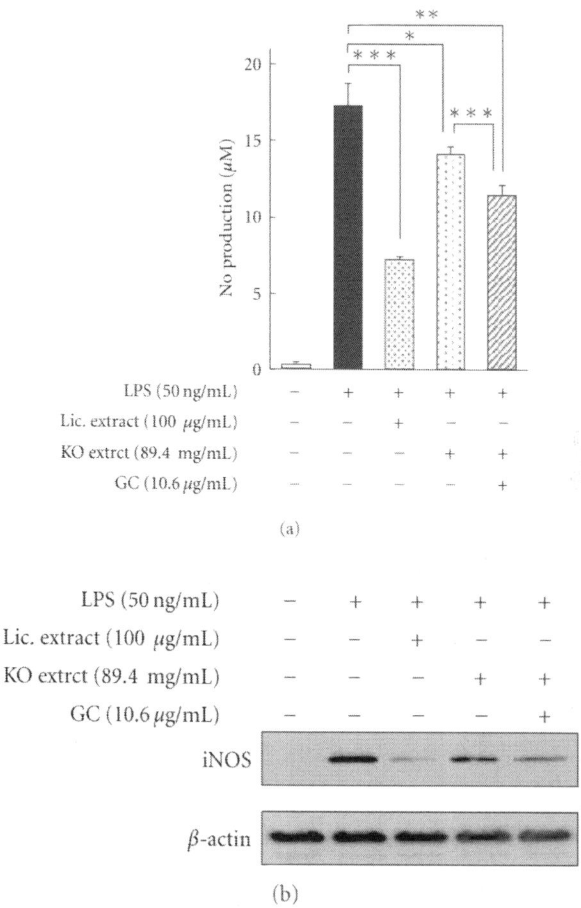

Figure 8. Effects of glycyrrhizin-knockout extract and the combination of glycyrrhizin-knockout extract and glycyrrhizin on LPS-induced NO production (a) and iNOS protein expression (b). $^*P < 0.05$, $^{**}P < 0.01$, $^{***}P < 0.001$.

CONCLUSIONS

Two unique applications using MAb, eastern blotting and knockout extract have been introduced in this paper. The eastern blotting method has great potential applications for the wide range of natural products, especially glycosides like ginsenosides of ginseng and glycyrrhizin in licorice. When two kinds of MAbs, against ginsenoside Rb_1 and Rg_1 can be used, the double staining that enhanced the separate staining of ginsenosides having protopanaxatriol or protopanaxadiol in a molecule, occurred. The staining color can be used to monitor the pharmacological activity suggesting that the purple spots contain protopanaxatriol as an aglycone indicating central nervous system (CNS) stimulatory activity [46]. On the other hand, the blue color indicates ginsenosides having protopanaxadiol which possess a depression effect on the CNS [37]. Furthermore, the Rf value of ginsenosides suggests the number of sugars attached to the aglycone. Both evidences make it possible to confirm which aglycone is attached and how many sugars are combined with the aglycone.

We demonstrated that knockout extract prepared by anti-natural compound-specific MAb immunoaffinity column is a useful approach for determination of potential function of natural compound on *in vitro* and *in vivo* assays. The pharmacological analysis by knockout extract might be directly applicable to both other crude extracts and various TCMs to clarify the real pharmacologically active components.

A ginsenoside Rb_1 knockout extract can be prepared using an immunoaffinity column conjugated with anti-ginsenoside Rb_1 MAb. In this extract, all compounds except only ginsenoside Rb_1 are contained. Furthermore, glycyrrhizin-knockout extract prepared from licorice crude extract using an immunoaffinity column conjugated with anti-glycyrrhizin MAb. These knockout extracts may be able to support the pharmacological investigation for finding out a really active component in a crude drug and/ or TCM. In fact, addition of glycyrrhizin to glycyrrhizin-knockout extract could improve the inhibition of iNOS expression resulting that glycyrrhizin may exert combinational inhibition of iNOS expression when coexisting with the other constituents contained in licorice extract.

REFERENCES

1. Sakata R, Shoyama Y, Murakami H. Production of monoclonal antibodies and enzyme immunoassay for typical adenylate cyclase activator, forskolin. *Cytotechnology*. 1994;16(2):101–108.
2. Ishiyama M, Shoyama Y, Murakami H, Shinohara H. Production of monoclonal antibodies and development of an ELISA for solamargine. *Cytotechnology*. 1996;18(3):153–158.
3. Xuan L, Tanaka H, Xu Y, Shoyama Y. Preparation of monoclonal antibody against crocin and its characterization. *Cytotechnology*. 1999;29(1):65–70.
4. Tanaka H, Goto Y, Shoyama Y. Monoclonal antibody based enzyme immunoassay for marihuana (cannabinoid) compounds. *Journal of Immunoassay*. 1996;17(4):321–342.
5. Shoyama Y, Fukada T, Murakami H. Production of monoclonal antibodies and ELISA for thebaine and codeine. *Cytotechnology*. 1995;19(1):55–61.
6. Tanaka H, Fukuda N, Shoyama Y. Formation of monoclonal antibody against a major ginseng component, ginsenoside Rb1 and its characterization. *Cytotechnology*. 1999;29(2):115–120.
7. Fukuda N, Tanaka H, Shoyama Y. Formation of monoclonal antibody against a major ginseng component, ginsenoside Rg1 and its characterization: monoclonal antibody for a ginseng saponin. *Cytotechnology*. 2000;34(3):197–204.
8. Kim JS, Tanaka H, Shoyama Y. Immunoquantitative analysis for berberine and its related compounds using monoclonal antibodies in herbal medicines. *Analyst*. 2004;129(1):87–91.
9. Morinaga O, Nakajima S, Tanaka H, Shoyama Y. Production of monoclonal antibodies against a major purgative component, sennoside B, their characterization and use in ELISA. *Analyst*. 2001;126(8):1372–1376.
10. Lu Z, Morinaga O, Tanaka H, Shoyama Y. A quantitative ELISA using monoclonal antibody to survey paeoniflorin and albiflorin in crude drugs and traditional Chinese herbal medicines. *Biological and Pharmaceutical Bulletin*. 2003;26(6):862–866.
11. Tanaka H, Shoyama Y. Formation of a monoclonal antibody against glycyrrhizin and development of an ELISA. *Biological and Pharmaceutical Bulletin*. 1998;21(12):1391–1393.

12. Shan S, Tanaka H, Shoyama Y. Enzyme-linked immunosorbent assay for glycyrrhizin using anti-glycyrrhizin monoclonal antibody and an eastern blotting technique for glucuronides of glycyrrhetic acid. *Analytical Chemistry*. 2001;73(24):5784–5790.

13. Loungratana P, Tanaka H, Shoyama Y. Production of monoclonal antibody against ginkgolic acids in Ginkgo biloba Linn. *American Journal of Chinese Medicine*. 2004;32(1):33–48.

14. Katsumi K, Edakuni K, Morinaga O, Tanaka H, Shoyama Y. An enzyme-linked immunosorbent assay for aconitine-type alkaloids using an anti-aconitine monoclonal antibody. *Analytica Chimica Acta*. 2008;616(1):109–114.

15. Kido K, Morinaga O, Shoyama Y, Tanaka H. Quick analysis of baicalin in Scutellariae Radix by enzyme-linked immunosorbent assay using a monoclonal antibody. *Talanta*. 2008;77(1):346–350.

16. Tanaka H, Fukuda N, Shoyama Y. Eastern blotting and immunoaffinity concentration using monoclonal antibody for ginseng saponins in the field of traditional Chinese medicines. *Journal of Agricultural and Food Chemistry*. 2007;55(10):3783–3787.

17. Wengatz I, Schmid R-D, Kreissig S, et al. Determination of the hapten density of immuno-conjugates by matrix-assisted UV laser desorption/ionization mass spectrometry. *Analytical Letters*. 1992;25:1983–1997.

18. Shoyama Y, Fukada T, Tanaka T, Kusai A, Nojima K. Direct determination of opium alkaloid-bovine serum albumin conjugate by matrix-assisted laser desorption/ionization mass spectrometry. *Biological and Pharmaceutical Bulletin*. 1993;16(10):1051–1053.

19. Shoyama Y, Sakata R, Isobe R, Murakami H, Kusai A, Nojima K. Direct determination of forskolin-bovine serum albumin conjugate by matrx-assisted laser desorption ionization mass spectrometry. *Organic Mass Spectrometry*. 1993;28:987–988.

20. Goto Y, Shima Y, Morimoto S, et al. Determination of tetrahydrocannabinolic acid-carrier protein conjugate by matrix-assisted laster desorption/Ionization mass spectrometry and antibody formation. *Organic Mass Spectrometry*. 1994;29:668–671.

21. The Society for Korean Ginseng. In *Understanding of Korean Ginseng*, The Society for Korean Ginseng, Seoul, Korea, 1995.

22. Kitagawa I, Taniyama T, Yoshikawa M, Ikenishi Y, Nakagawa Y. Chemical studies on crude drug processing. VI. (1)) Chemical structures

of malonyl-Ginsenosides Rb1, Rb2, Rc, and Rd isolated from the root of Panax ginseng C.A. Meyer. *Chemical and Pharmaceutical Bulletin*. 1989;37(11):2961–2970.

23. Morinaga O, Tanaka H, Shoyama Y. Detection and quantification of ginsenoside Re in ginseng samples by a chromatographic immunostaining method using monoclonal antibody against ginsenoside Re. *Journal of Chromatography B*. 2006;830(1):100–104.
24. Tanaka H, Putalun W, Tsuzaki C, Shoyama Y. A simple determination of steroidal alkaloid glycosides by thin-layer chromatography immunostaining using monoclonal antibody against solamargine. *FEBS Letters*. 1997;404(2-3):279–282.
25. Fukuda N, Shan S, Tanaka H, Shoyama Y. New staining methodology: eastern blotting for glycosides in the field of Kampo medicines. *Journal of Natural Medicines*. 2006;60(1):21–27.
26. Yokota S, Onohara Y, Shoyama Y. Immunofluorescence and immunoelectron microscopic localization of medicinal substance, Rb1, in several plant parts of Panax ginseng. *Current Drug Discovery Technologies*. 2011;8(1):51–59.
27. Morinaga O, Zhu S, Tanaka H, Shoyama Y. Visual detection of saikosaponins by on-membrane immunoassay and estimation of traditional Chinese medicines containing Bupleuri radix. *Biochemical and Biophysical Research Communications*. 2006;346(3):687–692.
28. Fukuda N, Tanaka H, Shoyama Y. Isolation of the pharmacologically active saponin ginsenoside Rb1 from ginseng by immunoaffinity column chromatography. *Journal of Natural Products*. 2000;63(2):283–285.
29. Yanagihara H, Sakata R, Minami H, Tanaka H, Shoyama Y, Murakami H. Immunoaffinity column chromatography against forskolin using an anti-forskolin monoclonal antibody and its application. *Analytica Chimica Acta*. 1996;335(1-2):63–70.
30. Putalun W, Tanaka H, Shoyama Y. Rapid separation of solasodine glycosides by an immunoaffinity column using anti-solamargine monoclonal antibody. *Cytotechnology*. 1999;31(1-2):151–156.
31. Tanaka H, Fukuda N, Yahara S, Isoda S, Yuan CS, Shoyama Y. Isolation of ginsenoside Rb1 from Kalopanax pictus by eastern blotting using anti-ginsenoside Rb1 monoclonal antibody. *Phytotherapy Research*. 2005;19(3):255–258.

32. Amagaya S, Sugishita E, Ogihara Y. Comparative studies of the stereoisomers of glycyrrhetinic acid on anti-inflammatory activities. *Journal of Pharmacobio-Dynamics*. 1984;7(12):923–928.
33. Doll R, Hill I-D. Triterpenoid liquorice compound in gastric and duodenal ulcer. *The Lancet*. 1962;280(7266):1166–1167.
34. Wang ZY, Nixon DW. Licorice and cancer. *Nutrition and Cancer*. 2001;39(1):1–11.
35. Kuroyanagi T, Saito M. Effect of prednisolone and glycyrrhizin on passive transfer in experimental allergic encephalomyelitis. *Japanese Journal of Allergology*. 1966;15(2):67–74.
36. Nakamura T, Fujii T, Ichihara A. Enzyme leakage due to change of membrane permeability of primary cultured rat hepatocytes treated with various hepatotoxins and its prevention by glycyrrhizin. *Cell Biology and Toxicology*. 1985;1(4):285–295.
37. Asl MN, Hosseinzadeh H. Review of pharmacological effects of glycyrrhiza sp. and its bioactive compounds. *Phytotherapy Research*. 2008;22(6):709–724.
38. Xu J, Tanaka H, Shoyama Y. One-step immunochromatographic separation and ELISA quantification of glycyrrhizin from traditional Chinese medicines. *Journal of Chromatography B*. 2007;850(1-2):53–58.
39. Wang C-Z, Shoyama Y. Herbal medicine: identification, and evaluation strategies. In: Yuan C-S, Bieber E-J, Bauer B-A, editors. *Text Book of Complementary and Alternative Medicine*. 2nd edition. Minnesota, Minn, USA: Informa Healthcare; 2006. pp. 51–70.
40. Moncada S, Palmer RMJ, Higgs EA. Nitric oxide: physiology, pathophysiology, and pharmacology. *Pharmacological Reviews*. 1991;43(2):109–142.
41. Nathan C, Xie QW. Regulation of biosynthesis of nitric oxide. *Journal of Biological Chemistry*. 1994;269(19):13725–13728.
42. Naseem KM. The role of nitric oxide in cardiovascular diseases. *Molecular Aspects of Medicine*. 2005;26(1-2):33–65.
43. Guzik TJ, Korbut R, Adamek-Guzik T. Nitric oxide and superoxide in inflammation and immune regulation. *Journal of Physiology and Pharmacology*. 2003;54(4):469–487.

44. Abramson SB, Amin AR, Clancy RM, Attur M. The role of nitric oxide in tissue destruction. *Best Practice and Research: Clinical Rheumatology*. 2001;15(5):831–845.
45. Uto T, Morinaga O, Tanaka H, Shoyama Y. Analysis of the synergistic effect of glycyrrhizin and other constituents in licorice extract on lipopolysaccharide-induced nitric oxide production using knock-out extract. *Biochemical and Biophysical Research Communications*. 2012;417(1):473–478.
46. Hashimoto K, Satoh K, Murata P, et al. Components of *Panax* ginseng that improve accelerated small intestinal transit. *Journal of Ethnopharmacology*. 2003;84(1):115–119.

7

Evaluating Strategies to Normalise Biological Replicates of Western Blot Data

Andrea Degasperi[1], Marc R. Birtwistle[2], Natalia Volinsky[1], Jens Rauch[1], Walter Kolch[1,3,4], Boris N. Kholodenko[1,3,4]

[1] Systems Biology Ireland, University College Dublin, Dublin, Republic of Ireland
[2] Department of Pharmacology and Systems Therapeutics, Icahn School of Medicine at Mount Sinai, New York, New York, United States of America
[3] Conway Institute of Biomolecular & Biomedical Research, University College Dublin, Dublin, Republic of Ireland
[4] School of Medicine and Medical Science, University College Dublin, Dublin, Republic of Ireland

ABSTRACT

Western blot data are widely used in quantitative applications such as statistical testing and mathematical modelling. To ensure accurate quantitation and comparability between experiments, Western blot replicates must be normalised, but it is unclear how the available methods affect statistical properties of the data. Here we evaluate three commonly used

Citation: Evaluating Strategies to Normalise Biological Replicates of Western Blot Data. Degasperi A, Birtwistle MR, Volinsky N, Rauch J, Kolch W, et al. (2014). PLoS ONE 9(1): e87293. doi:10.1371/journal.pone.0087293

normalisation strategies: (i) by fixed normalisation point or control; (ii) by sum of all data points in a replicate; and (iii) by optimal alignment of the replicates. We consider how these different strategies affect the coefficient of variation (CV) and the results of hypothesis testing with the normalised data. Normalisation by fixed point tends to increase the mean CV of normalised data in a manner that naturally depends on the choice of the normalisation point. Thus, in the context of hypothesis testing, normalisation by fixed point reduces false positives and increases false negatives. Analysis of published experimental data shows that choosing normalisation points with low quantified intensities results in a high normalised data CV and should thus be avoided. Normalisation by sum or by optimal alignment redistributes the raw data uncertainty in a mean-dependent manner, reducing the CV of high intensity points and increasing the CV of low intensity points. This causes the effect of normalisations by sum or optimal alignment on hypothesis testing to depend on the mean of the data tested; for high intensity points, false positives are increased and false negatives are decreased, while for low intensity points, false positives are decreased and false negatives are increased. These results will aid users of Western blotting to choose a suitable normalisation strategy and also understand the implications of this normalisation for subsequent hypothesis testing.

INTRODUCTION

Western blotting or protein immunoblotting, was introduced at the end of the 1970s to enable the detection of specific proteins [1], [2]. Although originally a qualitative or at best a semi-quantitative method, with the rise of computational systems biology [3], Western blotting has become increasingly important for fully quantitative applications. Two main applications are the parameterisation and validation of mathematical models of biological systems [4] and the testing of statistical significance between two or more experimental conditions or treatments [5].

Although technical aspects of Western blotting have improved over the years, for example by extending the linear range of detection [6], it is not yet clear how much quantitative information can be obtained and in which settings. Here we investigate the quantitative use of Western blotting, to determine its applicability and limits depending on the detection method and the data normalisation strategy used to quantitatively compare biological replicates of the same experimental conditions.

A requirement for the quantitative use of Western blot data is the linearity between quantified intensities and protein amounts. To detect and correct non-linearity, the authors in [7] suggest to use hyperbolic calibration curves to interpolate the correct relative concentration of the proteins of interest. These are dilution curves that need to be treated simultaneously to samples of interest, and in most situations constructing these dilution curves is not practical. Because this method is highly labour consuming and is not a laboratory common practice, we do not consider this approach in this paper. Nonetheless we investigate linearity in our Results section, where we quantify the extent of the linear range in the case of two detection systems: enhanced chemiluminescence (ECL) with X-ray film and ECL with charge coupled device (CCD) imager.

Although the topic of data normalisation has been widely explored in the context of microarrays [8], it has not yet been fully investigated in the context of Western blotting. For microarrays, such as single channel oligonucleotide arrays, the issue of data normalisation arises naturally when expression indices, obtained from gene probe sets intensities, need to be compared across different arrays, for example to identify differentially expressed genes [9]. In order to compare arrays quantitatively, several normalisation strategies have been proposed, where expression indices or intensities are scaled or transformed depending on the assumptions underlying each strategy. For example, assuming that the total amount of sample RNA is constant across arrays, the intensities are scaled such that the sum or the average of all intensities is equal across arrays (scaling methods [9]). Alternatively, assuming that the distribution of the intensities is conserved across arrays, the data is transformed such that the quantile-quantile plot of the intensities of the arrays approaches a straight line (quantile normalisation [9]). Or again, assuming that there is a set of genes whose expression index does not change across arrays, such as a set of housekeeping genes, this set can be used as reference (invariant set normalisation [10]).

In the case of Western blotting, usually a single protein is measured and a limited number of experimental conditions is on the same blot and is detected at the same time; a situation in stark contrast to microarrays where thousands of gene expression measurements are obtained for potentially many more conditions than is typically done by Western blotting. The data or measurements from a Western blot are obtained by dividing quantified intensities (optical densities – OD) by the intensities of appropriate reference proteins, e.g. housekeeping proteins, from the same samples. This procedure adjusts the intensities with respect to small variations in the number of cells and loading across samples within the same blot [11]–[13]. The need to normalise the data arises when comparing the results from biological replicates of the same experiment, for example to obtain statistical evidence that different conditions induce different protein amounts. We classify Western blot normalisations into three categories. The first and most widely used normalisation method is normalisation by fixed point (Figure 1A). It divides the data of a replicate by the measurement of a single condition, often referred to as control. It should be noted that although this shares similarities with the invariant set normalisation in the context of microarrays, the assumption that the reference condition is constant is not used and is in practice not satisfied. Thus the biological variability of the reference condition influences the variability of the normalised data. The second normalisation category we consider is normalisation by sum (Figure 1B), where the data on a blot is divided by the sum of the data on the same blot [14], or equivalently the data is scaled such that the average is the same across blots [15], [16]. It should be noted that in contrast with the analogous normalisation used in microarrays (scaling methods), in this case the sum is not assumed to be a constant. The biological variability of the sum and its dependency on the individual measurements might influence the variability of the normalised data. Most importantly, it is possible to compare two blots only if they present exactly the same conditions, using different lysates derived from cells cultured and treated in the same way. Yet, this condition is regularly met when producing a biological replicate. In our statistical formalisation we address the problem of characterising how the choice of reference (fixed point or sum) influences the normalised data. As third category we consider normalisation by optimal alignment (Figure 1C), where data from replicates are aligned using optimisation algorithms to minimise the uncertainty of the normalised data. Examples of this normalisation minimise either the sum of the squared differences between the replicates of each data point [17] or the coefficient of variation (CV) of the normalised data [18]. The assumption behind this method is that the measurements across replicates should preserve an overall trend.

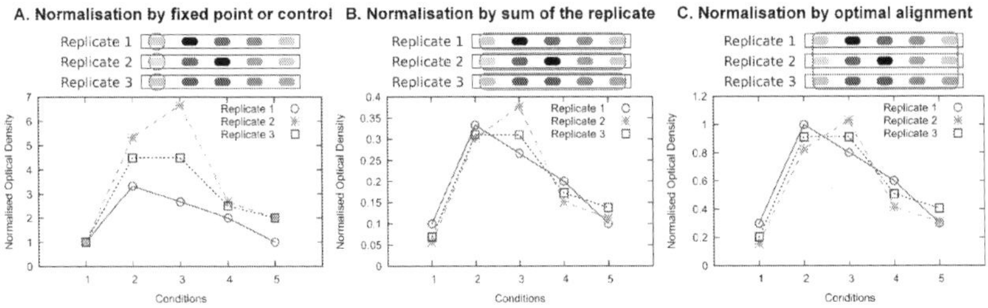

Figure 1. Normalisations of Western blot replicates in the literature.

We divide the normalisations found in literature into three categories: (A) normalisation by fixed normalisation point or control; (B) normalisation by sum of the replicate; (C) normalisation by optimal alignment. For illustration purposes we do not use actual Western blot data. Each normalisation is presented using three cartoon Western blots, representing three replicates, and highlighting with red circles the data points used in the normalisation procedure. The graphs show the normalised data, where the points belonging to the same replicate are connected with lines.

To avoid the need for data normalisation, approaches for the absolute quantification of protein concentrations using Western blotting have been investigated [7], [19]. However, these methods are not widespread mainly due to increased experimental effort, in particular the need for purified proteins as standards. It is also possible to obtain replicates of lysates that are directly comparable by means of multi-strip Western blotting [20], where replicates are cut from different gels and blotted on the same membrane. However, multi-strip Western blots are typically used to compare more conditions on the same membrane, rather than replicates.

The quantitation of Western blots has also been the subject of theoretical investigations. In [21] the authors use a large amount of data to identify a suitable error model for Western blot data. Using the error model, they dissect the different sources of error, concluding that the main sources of variability are multiplicative and so log-normally distributed. Additionally, by removing the sources of error, they reduce the variability in the data significantly. This work is based on error models for microarray data [22], and is applicable only when a large amount of data is available. In [19], the authors suggest that technical errors can be reduced using a randomisation of time courses on a gel and smoothing the data using spline regression.

In this paper, first we discuss the problem of linearity between protein concentrations and quantified optical densities, which is a fundamental prerequisite to use Western blot data quantitatively in the absence of hard-to-obtain calibration curves. Second, we investigate how the choice of the normalisation strategy affects the normalised data. In particular, we evaluate the normalisations in terms of their ability to reduce variability in the data and of how they affect statistical decisions.

MATERIALS AND METHODS

Sample Preparation

The MCF-7 cell line was maintained under standard conditions in Dulbecco's modified Eagle's medium supplemented with 10% foetal bovine serum. Cells were washed with ice cold phosphate buffered Saline and lysed in RIPA buffer (1% NP-40, 0.1% SDS, 0.5% Sodium deoxycholate, 50 mM Tris pH 7.5, 150 mM NaCl) supplemented with protease and phosphatase inhibitor cocktails (Sigma Aldrich) and protein concentration was quantitated by BCA protein assay (Invitrogen). Purified BSA (Applichem) was dissolved in RIPA buffer. Cell lysates and a BSA sample were serially diluted 1:2 and run on SDS-PAGE using a standard protocol. Proteins were transferred to the PVDF (for ECL based detection) or Nitrocellulose (for LI-COR based proteins detection) membranes. Membranes were blocked with blocking solution (11500694001, Roche) for BSA detection or 5% skimmed milk for rest of the membranes. For Western blotting ERK (M-5670, Sigma Aldrich), mTOR (2972, Cell Signaling Technology), RSK1 (sc-231, Santa Cruz) and BSA (sc-50528, Santa Cruz) antibodies were used. Anti-rabbit HRP-conjugated (Cell Signaling Technology) or anti-Rabbit IR 800 (LI-COR) secondary antibodies were used for ECL or LI-COR protein detection systems, respectively. Signal was detected by standard X-ray films (Fuji), CCD camera (Advanced Molecular Vision) or LI-COR scanner.

Image Acquisition and Densitometry

Several exposure times were tested for both ECL with film and ECL with CCD imager. In the case of the CCD imager we could choose the longest exposure that presented no signal saturation (overexposure), as detected by the software used in combination with the imager. In the case of X-ray film, we used our experience to select the films that had a good compromise between number of bands visible and the least possible exposure time. Films were digitalised using a high resolution CCD scanner (EPSON Perfection v750 Pro) without additional image corrections that could alter the linearity, such as automatic gain control [23]. Densitometry analysis was performed using the ImageJ Gel Analysis tool, where gel background was also removed individually for each band.

Statistical Analysis of the Dilution Experiments

To assess the linearity of the dilution experiments we used linear regression and computed the coefficient of determination R^2 using Microsoft Excel software. Briefly, the closer the coefficient of determination is to 1, the more the linear model is appropriate to represent the data. For each detection method we tested different dilution ranges by determining R^2 using the first n visible bands, i.e. starting from the least intense band that could be detected. For example, to test the linearity of the five dilutions range, reflecting a 32 fold difference, we computed the linear regression of the first six (n=6) visible bands of a dilution curve and computed R^2. We then computed R^2 independently for three replicates and obtained mean and standard error. Using this approach we could compare coefficients of determination for specific dilution ranges across different detection methods.

Theoretical Analysis of Normalisation Procedures

The theoretical analysis was performed developing dedicated scripts in the R language for statistical computing and implementing dedicated C++ programs. While R was used mainly to compute the results inferred from Western blot data, C++ programs were used to compute the results in the theoretical scenarios. To obtain samples from log-normal distributions, we computed samples from normal distributions using the Box-Muller method [24] and then exponentiated these samples with base e. Mean and variance of these normal distributions were calculated so that the mean and variance of the log-normal distributions were as desired. Combining samples from log-normal distributions we could obtain samples from the distributions of the normalised data, as defined in the text (Equations (2), (4) and (8)). Because the log-normal distributions were defined *a priori*, we could then estimate false positives and false negatives results by using t-tests as described in the corresponding figure legends. Briefly, a false positive is defined as a t-test result that yields a p-value lower than 0.05 when testing samples from two distributions that we defined as identical, while a false negative is defined as a t-test result that yields a p-value greater than 0.05 when testing samples from two distributions that we defined as different. Source files are available upon request from the corresponding author.

RESULTS

Linearity between Protein Concentration and Quantified Optical Densities

In the absence of a calibration curve, a pre-requisite for obtaining quantitative Western blot data is a linear relationship between the amount of analyte and the measured intensity. To evaluate the extent of the linear range in commonly used detection methods, we prepared two 12 step 2-fold dilution series spanning a 2048-fold concentration range (three independent experiments). One sample series contained isolated Bovine Serum Albumin (BSA) while the other MCF-7 cell lysate. We used the first dilution series to quantify BSA and the second to quantify proteins across a mass range; they included Extracellular signal Regulated Kinases 1 and 2 (ERK1/2), ca. 40 kDa, Ribosomal protein S6 Kinase alpha-1 (RSK1), ca. 80 kDa, and Mammalian Target Of Rapamycin (mTOR), ca. 290 kDa. Proteins were detected using two detection systems: ECL with X-ray film and ECL with a CCD imager. Representative experiments with corresponding quantifications can be found in Figure 2 (BSA) and in Figures S1 (ERK), S2 (RSK1) and S3 (mTOR).

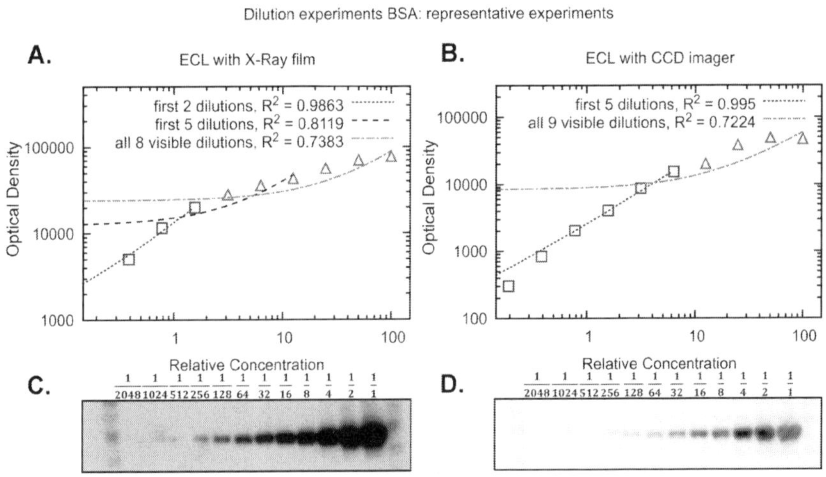

Figure 2. Signal linearity obtained by different Western blot detection systems.

Representative experiments of Western blots containing 2-fold serial dilution of BSA. Shown are the representative results from 3 independent experiments. BSA was detected by (A,C) ECL with X-ray film and (B,D) ECL with CCD imager. Blue squares indicate data points that are linear, while red triangles indicate data points outside the linear range of detection. To highlight linear and non-linear data we use linear trend lines, reporting the coefficient of determination R^2. In (A,B) data are in log-log scale to improve visualisation.

In order to identify the linear range of a dilution curve we used linear regression for an increasing number of data points, starting from the first detectable and least intense band of a curve. For each regression we then computed the coefficient of determination R^2, which indicates if the linear regression is a good model for the portion of the curve considered. The closer R^2 is to 1, the more linear the data is. After computing R^2 for each of three replicates we obtained mean and standard error.

As expected, our results show that both ECL with X-ray film and with CCD imager have a limited linear range. For example, the full dilution curve of BSA detected with ECL with CCD imager has an R^2 of 0.788 ± 0.035.

Interestingly, we found that the linear range of ECL with CCD imager spans about five dilutions (32 fold). In particular, for all four proteins considered we obtained a significant reduction in R^2 when we consider six dilutions (64 folds), with respect to five dilutions. For example, in the case of BSA, the coefficient of determination for five dilutions is 0.994 ± 0.001, while for six dilutions it is reduced to 0.952 ± 0.011, while for ERK the reduction is from 0.993 ± 0.003 to 0.994 ± 0.01.

ECL with X-ray film presented a smaller linear range than with CCD imager. In particular, for the five dilutions range where CCD imager is linear, X-ray film yields a sensibly lower R^2. For example, for the BSA dilution this is reduced to 0.830 ± 0.023, while for ERK to 0.732 ± 0.025. Linearity in the case of ECL with X-ray film seems to hold only for two or three dilutions, i.e. four or eight fold.

The difference between the two systems based on ECL is most likely due to saturation of the X-ray film by high intensity samples while trying to detect also the lowest intensity samples. This limitation can be avoided using a CCD imager, which uses a computerised image acquisition system, and is able to detect low intensity signals without high intensity signals becoming saturated as quickly as with film. Because we were able to avoid this overexposure, the non-linearity observed using the CCD imager (Fig. 3B) is likely due to antibody interactions, as suggested in [7].

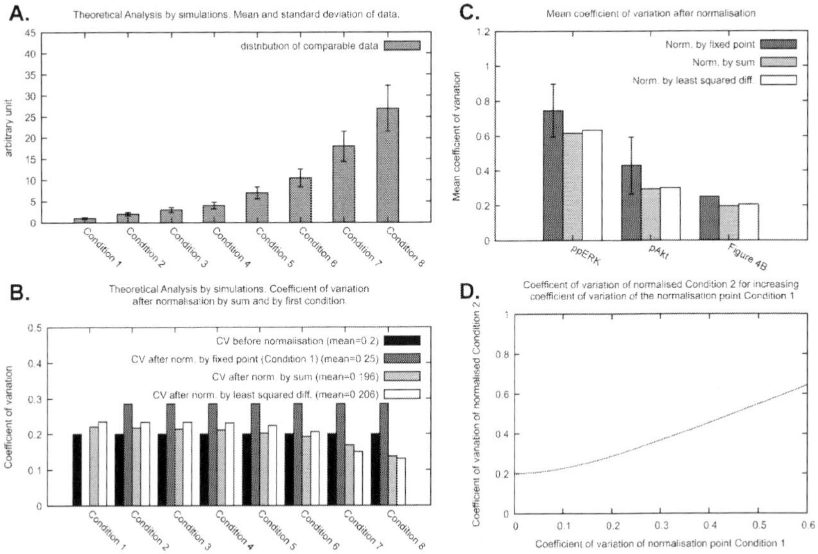

Figure 3. Effect of the normalisation on the CV of the normalised data.

(A) Distribution of the data in a simulated scenario. In our theoretical analysis of the effects of the normalisation on the variability of the normalised data we consider a distribution of the response to eight conditions. We use log-normal distributions with CV 0.2 and mean of the response to the conditions from 1 to 8 as 1, 2, 3, 4, 7, 10.5, 18, 27. (B) CVs are shown for the distribution of the simulated data before normalisation, after normalisation by first condition, after normalisation by sum of all data points in a replicate and after normalisation by least squared differences. The mean CV is computed as the average across the eight conditions. (C) Data from Figure S3 of [25] (Figure S5 in this publication) were normalised using different normalisation strategies and the mean CV of the resulting normalised data is shown. As the mean CV obtained by the normalisation by fixed point depends on the choice of normalisation point, we report the mean and standard deviation obtained. We also report the mean CV obtained using ppERK and pAkt data and we compare them with the theoretical results of Figure 3B. (D) Before normalisation, the response to Condition 2 has a CV of 0.2, as shown in Figure 3A. Condition 2 is then normalised by fixed point, with Condition 1 as normalisation point. Here we show how the CV of normalised Condition 2 changes for increasing CV of the normalisation point Condition 1.

Finally, we investigated the extent of the linear range when using secondary fluorescent antibodies of LI-COR to detect BSA and ERK (Figure S4). Results were comparable to what we described above for ECL with CCD imager.

In conclusion, the use of ECL with X-ray film for quantitative Western blotting should be limited to the case in which the intensities vary experimentally not more than 4 to 8 fold. ECL with CCD imager or secondary fluorescent antibody presents a wider linear range of about 32 fold.

Formalisation of Normalisation Strategies for Western Blot Data

We now move the focus to the evaluation of the normalisation strategies that we categorised in the introduction. In this section we introduce a formalisation, i.e. a mathematical description, of the normalisation strategies we investigate. Without loss of generality, consider the measurements of a single target (e.g. a protein) under different conditions or treatments (e.g. inhibitors, stimuli) on the same blot. Each data point d_i^j, is indexed by the condition $i \in I$ and the blot replicate number $j \in J$. The standard experimental setups described above dictate the following:

1. Data points on one blot are comparable to one another, even if they come from different gel strips [20]. That is, d_i^j with different i but with the same j are directly comparable;
2. Data points on two different blots are not comparable. That is, d_i^j with different j are not directly comparable. Normalisation must be employed to enable direct comparison across replicates.

Given the linearity conditions explored above are met and that the data points d_i^j are samples from random variables D_i^j, we have:

$$D_i^j = \alpha_j \cdot [S]_i \qquad \forall i \in I, \forall j \in J \tag{1}$$

where $[S]_i$ is the concentration of the protein of interest S in conditioni i, and α_j is the constant of proportionality of replicate j. Note that if the replicates had been blotted on the same membrane using multi-strip Western blotting, all the data would be comparable with $\alpha_j = \alpha$ for all $j \in J$.

The distribution of $[S]_i$ depends on the mean of the concentration of S across a large number of cells in the lysate, on the biological variability, and on the technical error. We will now introduce the formalisation of the normalisation by fixed point or control, by sum of the replicate, and by optimal alignment of the replicates.

Normalisation by a Fixed Normalisation Point or Control

In the normalisation by fixed normalisation point or control a band on the blot common to all replicates is chosen to be the normalisation point, and the data from a replicate are divided by the value of the normalisation point. In general, this normalisation can be applied choosing any band that is present on all replicates that need to be compared. The term "control" usually indicates the band of the untreated or neutral condition. Formally, this normalisation is a data transformation where the data points d_i^j are substituted by the normalised data points $\tilde{d}^j_{i,np}$ defined as:

$$\tilde{d}^j_{i,np} = \frac{d_i^j}{d_{np}^j} \qquad \forall i \in I, \forall j \in J \tag{2}$$

Where the index np indicates a chosen normalisation point, which is an experimental condition all normalised data become relative to. In terms of random variables, the $\tilde{d}^j_{i,np}$ are samples from the random variables $\tilde{D}^j_{np,np}$ defined as:

$$\tilde{D}_{i,np} = \frac{D_i^j}{D_{np}^j} = \frac{\alpha_j \cdot [S]_i}{\alpha_j \cdot [S]_{np}} = \frac{[S]_i}{[S]_{np}} \qquad \forall i \in I, \forall j \in J \tag{3}$$

Notice that the random variable $\tilde{D}^j_{np,np}$ assumes value 1 with probability 1. Most importantly, the $\tilde{d}^j_{i,np}$ are now comparable across replicates j, because the constants of proportionality α_j are all cancelled out. Additionally, all normalised data is dependent on the distribution of the normalisation point. In the following sections we will show how this dependency influences the variability of the normalised data and we will investigate how to choose a normalisation point.

Normalisation by Sum of all Data Points in a Replicate

In the normalisation by sum, each data point on a replicate is divided by the sum of the values of all data points in that replicate. This way the data in each replicate becomes relative to this sum. It is important to ensure consistency of the sum across replicates, that is exactly the same conditions need to be part of the sum. This ensures that each data point is divided by a sample that comes from the same random variable. For example, in the presence of missing values, data points to be summed are chosen so that no replicate of the corresponding condition has a missing value. Without loss of generality, in this section we give a formalisation of normalisation by sum where we do not consider missing values.

In this normalisation, data points d_i^j are divided by the sum of all data points in a replicate j. Formally, the normalised data $\tilde{d}_i^j$ are defined as:

$$\tilde{d}_i^j = \frac{d_i^j}{\sum_{k \in I} d_k^j} \qquad \forall i \in I, \forall j \in J \tag{4}$$

In terms of random variables, the normalised data points $\tilde{d}_i^j$ are samples from the random variables $\tilde{D}_i$ defined as:

$$\tilde{D}_i = \frac{D_i^j}{\sum_{k \in I} D_k^j} = \frac{\alpha_j \cdot [S]_i}{\alpha_j \cdot \sum_{k \in I} [S]_k} = \frac{[S]_i}{\sum_{k \in I} [S]_k} \qquad \forall i \in I, \forall j \in J \tag{5}$$

Similar to the normalisation by fixed point, the constants of proportionality α_j cancel and comparable normalised data are obtained. Notice that the normalised data are dependent on the value of all the data points in a replicate. The effects of this dependency on the variability of the data and hypothesis testing are investigated in the following sections. We note that Eq. 5 may be obtained by formulating the normalisation as an optimisation problem.

Normalisation by Optimal Alignment of the Replicates

In the normalisation by optimal alignment, the objective is to scale the data by a scaling factor for each replicate, so that replicates are aligned, that is the distance between data across replicates is minimal. This procedure has the specific goal of minimising the variability of the normalised data. Moreover, different notions of distance can be used, yielding different definitions of objective functions. The objective functions formalise the distance between the data in the replicates and are parametric with respect to the scaling factors. Finding the minimum of an objective function implies identifying optimal scaling factors. Examples of objective functions are as the sum of the squared differences between replicates [17] or the mean CV of the normalised data [18]. In the following we give a formal definition of normalisation by optimal alignment, considering a specific definition of distance, i.e. the sum of squared differences between replicates.

In this normalisation, each replicate J is scaled by a factor β_j so that an optimal alignment of the replicates is achieved. It is necessary to avoid the trivial solution $\beta_j = 0$ for all $j \in J$, which can be done by introducing the constraint $\beta_1 = 1$ and estimating the remaining β_j. Here we consider the normalisation by least squared difference, defined as follows. Assuming we have |J| replicates and we want to align every replicate to the first replicate, the objective function for a least squared optimal alignment is as follows:

$$obj\left(\beta_2,\ldots,\beta_{|J|}\right) = \sum_{j \in J \setminus \{1\}} \sum_{i \in I} \left(\beta_j \cdot d_i^j - \beta_1 \cdot d_i^1\right)^2, \text{with } \beta_1 = 1$$

We minimise obj to find the optimal $\beta_2,\ldots,\beta_{|J|}$. The result of this optimisation can be computed analytically and yields:

$$\beta_j = \frac{\sum_{i \in I} (d_i^1 \cdot d_i^j)}{\sum_{i \in I} (d_i^j)^2} \qquad \forall j \in J \tag{7}$$

The normalised data $\tilde{d}_i^j$ are defined as follows:

$$\tilde{d}_i^j = d_i^j \cdot \beta_j = d_i^j \cdot \frac{\sum_{k \in I} (d_k^1 \cdot d_k^j)}{\sum_{k \in I} (d_k^j)^2} \qquad \forall i \in I, \forall j \in J \tag{8}$$

Notice that because of the definition of the β_j, a normalised data point depends on a combination of the value of the data in the same replicate and the value of the data in replicate 1. More complex normalisations by optimal alignment of the replicates, such as the normalisation by minimisation of the mean CV of the normalised data in [18], may present normalised data points that are dependent on the values of all the data. For illustration purposes we use here normalisation by least squared difference as a representative of the normalisations by optimal alignment.

Impact of Normalisation on Data Variability

A major aim of data normalisation is to make replicates suitable for quantitative comparison, while ensuring data integrity and avoiding adding uncertainty to the data. Here we show how different normalisation strategies affect the variability of the normalised data. We use the CV of the normalised data to compare the variability that results from applying the different normalisations.

For a theoretical investigation of how the choice of normalisation strategy affects the data, we use a simulated scenario. Suppose that the data of eight conditions or treatments is given as in Figure 3A. We chose a data distribution where the response to the treatments from one to eight has an increasing mean but the same CV of 0.2. In this and further analyses, we consider these distributions to be log-normal, because of the finding in [21] that the main sources of variability in Western blot data are multiplicative, and therefore log-normally distributed.

In Figure 3B we show how normalisation by fixed point, normalisation by sum and normalisation by least squared difference affect the CV of the eight conditions. To obtain these results we estimated the distributions associated with the random variables that we identified in Equation (3), Equation (5), using a sampling approach based on the Box Muller sampling method [24]. We chose Condition 1 as the normalisation point for the normalisation by fixed point. It should be noted that because every condition is distributed with the same CV, this choice is an invariant in our analysis. The mean of the normalisation point does not determine the CV of the normalised data (data not shown), while we will show below that the CV of the normalised data depends strictly on the CV of the normalisation point chosen and that in practice data points with low mean, i.e. low OD, usually present higher CV. Normalisation by fixed point induced an increase in the mean CV, from 0.2 to 0.25, while increasing equally the CV of the response to each condition, with the obvious exception of Condition 1. Normalisation by sum slightly reduced the mean CV, from 0.2 to 0.196, while the effect on the single responses to the conditions is a redistribution of the CV in a way that is dependent on the mean of the conditions. Conditions with high mean present a reduced CV, while conditions with low mean present an increased CV (albeit slightly). This redistribution is due to the fact that the distribution of the data normalised by sum is dependent on the distribution of all the conditions, as can be seen in Equation (5). Normalisation by least squared difference optimisation increased very slightly the mean CV, from

0.2 to 0.206, while the effect on the single responses to the conditions is a redistribution of the CV analogous to what is observed for the normalisation by sum.

In order to obtain evidence that support the above theoretical investigation, we apply the normalisations to our Western blot data published in Figure S3 of [25] and also available in Figure S7. These data are composed of two data sets, one with measurements of phosphorylated ERK (ppERK), and the other with measurements of phosphorylated Akt (pAkt). Each data set is composed of three replicates of multi-strip Western blots, where on each blot there are 70 conditions divided into seven time courses for two different cell lines. Each data set is thus composed of 210 data points of measurements of band intensities already divided by the intensity of the corresponding loading controls, which are total ERK and Akt, respectively. These blots were done using ECL for protein detection and CCD imaging for recording of band intensities. All measurements were detected avoiding overexposure, and as most of the measurements are within a limited dynamic range they are likely within a linear range of detection. Figure 3C illustrates the mean CV obtained for the two data sets after applying different normalisation strategies, and compares these results with the theoretical investigation of Figure 3B. The results obtained with the experimental data agree with the theoretical investigation. The mean CV of the normalised data is relatively low for the normalisation by sum and the normalisation by least squared difference, while higher, on average, for the normalisation by fixed point. In practice, the result of the normalisation by fixed point depends on the choice of the normalisation point, yielding normalised data with low and high variability depending on such choice. In the next section we investigate how to choose a normalisation point.

Low Intensity Data Points are Unsuitable Normalisation Points

In this section we investigate how, in the normalisation by fixed point, the choice of the normalisation point affects the variability of the data. In Figure 3D we illustrate how an increase in the CV of the normalisation point (Condition 1) induces a monotonic increase in the CV of the distribution of the normalised data (Condition 2), estimated using Equation (3). This result implies that the CV of the data normalised by fixed point is directly correlated with the CV of the particular condition used as normalisation point. This result also implies that a good choice for a normalisation point is a condition that presents a response with low CV, and hence low uncertainty. Although for non-comparable biological replicates it is impossible to pinpoint which data points have low variability, in the following we provide evidence that low protein band intensities usually yield normalised data with high variability.

For this analysis we again use the Western blot data in Figure S7. For both ppERK and pAkt data sets we use each data point as normalisation point and calculate the average CV of the other points resulting from this normalisation. The results are illustrated in Figure 4 where, using a regression by spline functions, we show that choosing low intensity bands as a normalisation point causes an increase in the mean CV of the normalised data. Additionally, because of the result in Figure 3D, we can infer that low intensity bands have usually a larger CV and thus a higher uncertainty. This is most likely due to the low signal-to-noise ratio or, in other words, due to the presence of background noise and the difficulty to separate this noise from low intensity measurements.

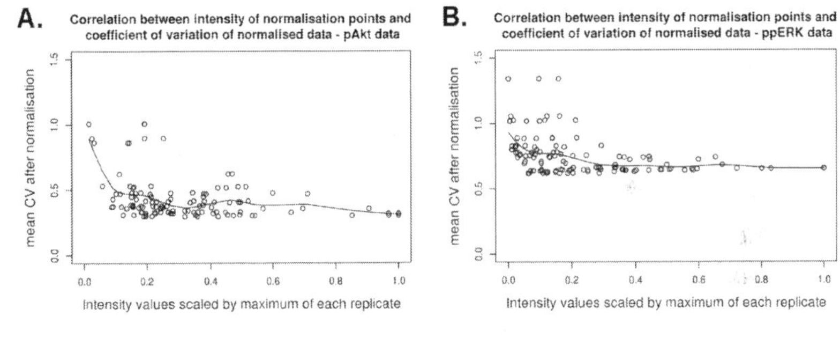

Figure 4. Correlation between the intensity of the normalisation points and the CV of the normalised data.

Using data from (A) phosphorylated Akt and (B) phosphorylated ERK from Figure S3 in [25] (Figure S5 in this publication) we tested every point on a blot as normalisation point. For each resulting normalisation we computed the average of the CV of the normalised data points, and plotted the value of each data point (scaled so that the maximum of each replicate is equal to 1) against the average CV obtained by normalising with the corresponding data point. The result shows how the intensities of each normalisation point chosen correlate with the variability of the normalised data.

In addition, we investigated whether the presence of data points that are outside the linear range of detection could also affect the choice of normalisation point. We performed an analysis analogous to what we described above for Figure 4 using the data from the three replicates of the dilution experiments illustrated in Figure 2 and Figures S1, S2 and S3.

We used the data of the ECL with CCD imager detection system, which all present non-linearity outside the 32 fold linear range. The results for proteins BSA, RSK1 and mTOR present similarities with what we have found for ppERK and pAkt in Figure 4, i.e. using lower intensity measurements as normalisation point induces larger CV. The result for ERK is shown in Figure S8 and indicates that in this case both high intensity and low intensity normalisation points induce high CV, while medium intensity measurements induce the smallest CV. This is most likely due to the fact that the hyperbolic part of the dilution curves, which is composed of high intensity bands, is not reproduced consistently across replicates. Thus, the variability of high intensity data that are outside the linear range can induce normalised data with large CV even when a high intensity measurement is used as normalisation point.

Impact of Normalisation on Statistical Testing

In this section we use a simulated scenario to investigate the effects of normalisation on the statistical testing applied to examine the significance of differences between protein bands detected by Western blotting. In particular, we test how normalisations influence the sensitivity and specificity of the two-tailed t-test [5], which is frequently used. In order to evaluate the sensitivity and the specificity, we estimate the percentage of false positives and false negatives by repeated data sampling.

It is standard practice to employ the t-test in spite of the fact that the actual distribution of Western blot data is unknown and theoretical investigation points toward a log-normal distribution [21], which is different from the normal distribution assumed in the test. Fortunately, the t-test is robust with respect to this violation of its assumptions [26], [27] having enabled its widespread application to Western blot data. As we want our analysis to be relevant for the practitioners we therefore comply with the established practice, and also use the t-test.

In Figures 5A,B, the row labelled "Before normalisation" illustrates the expected percentage of false positives and false negatives obtained applying the t-test to data distributed as described in the previous figures. Because we set a threshold p-value of 0.05 in the t-test, the percentage of false positive is about 5%, as expected. Variations to the percentage of false positives and false negatives should be attributed to the application of the normalisations.

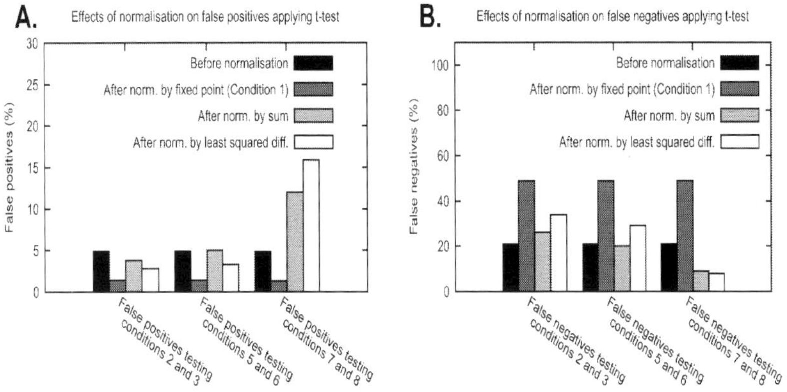

Figure 5. Effects of normalisation on false positives and false negatives when applying t-test for equality of the mean. (A) We consider responses to eight conditions with log-normal distributions with CV of 0.2 and means of the conditions from 1 to 8 equal to: 1, 2, 2, 4, 7, 7, 18, 18. A number n=5 of sampled replicates are obtained from these distributions and normalised using the normalisations above. Using these replicates before and after normalisation, conditions are tested using a two-tailed t-test with threshold p-value of 0.05. We repeat this procedure a large number of times and estimate the percentage of false positives. (B) In analogy with (A), we estimate the number of false negatives considering means of the conditions from 1 to 8 equal to: 1, 2, 3, 4, 7, 10.5, 18, 27. Notice that for a fair comparison, when testing two conditions, one has a mean that is always 2/3 the mean of the other, e.g. Condition 5 has mean 7 and Condition 6 has mean 10.5, with 7/10.5=2/3.

Normalisation by fixed point reduces the percentage of false positives, but greatly increases the percentage of false negatives, i.e. the specificity of the test is increased but the sensitivity is greatly reduced. This result is in agreement with our finding that the normalisation by fixed point increases the CV of the data. Thus, choosing this normalisation method will fail to identify some of the differences between data points. Normalisation by sum affects the percentage of false positives and false negatives in a way that is dependent on the mean of the response to the conditions tested. If relatively low values are tested, e.g. Conditions 2 and 3 in Figures 5A,B, the number of false positives decreases and the number of false negatives increases, while if relatively high values are tested, e.g. Conditions 7 and 8, the number of false positives increases and the number of false negatives decreases. Normalisation by least squared difference also affects false positives and false negatives depending on the magnitude of the data tested. Additionally, it seems that the normalisations by optimal alignment, such as by least squared difference, induce a stronger change in the sensitivity and specificity than the normalisation by sum. In general, normalisation by sum and by optimal alignment can introduce false positives when testing data with values relatively higher than the rest of the data set, reducing the specificity of the test.

DISCUSSION

In this paper we have investigated two issues that are important for the quantitative use of Western blot data, i.e. linearity of the detection system and the influence of data normalisations. Our results indicate that for quantitative Western blotting, if the measured intensities vary more than 4–8 fold, then the ECL detected by CCD imager system is preferable to ECL detected by X-ray film, as it yields a larger linear dynamic range. The linear range in the case of ECL in combination with CCD imager spans about 32 fold concentration change for four different proteins with different molecular mass, (Figure 2 and Figures S1, S2 and S3). When we tested fluorescent secondary antibodies detected by the LI-COR scanner, we found that a linear range is similar to the ECL with CCD imager detection method (Figure S4).

To better understand the mechanisms behind the normalisation of Western blot data, we use a formalisation based on statistical arguments of three normalisation strategies. Our findings reveal that the normalisation by fixed point introduces additional variability in the data (Figure 3B), and that conditions that induce responses with low CV are preferable normalisation points, because they induce a lower CV of the normalised data (Figure 3D). Although the CV of the response to specific conditions is in general not known, we provide evidence of whether low, medium or high intensity measurements have usually high or low CV. In particular we showed that low intensity measurements are usually inappropriate normalisation points (Figure 4). This is most likely due to the low signal-to-noise ratio and consequent high CV of low intensity measurements. Additionally, we showed that high intensity measurements that are outside the linear dynamic range are inappropriate normalisation points (Figure S8). Therefore, we suggest that for this type of normalisation the most appropriate normalisation points are data points with medium intensity measurements. Because the normalisation by fixed point increases the CV of the normalised data, this also has an impact on statistical testing. When applying a two-tailed t-test to the normalised data, we saw an increase in the specificity of the test and a strong decrease of the sensitivity (Figure 5A,B). While a high specificity is desirable, the decline in sensitivity increases the chances of overlooking significant differences between data points. In addition, if the normalisation point is not chosen carefully, the normalised data could present a high variability and it might become very difficult to detect when two conditions yield different results.

The normalisations by sum and by optimal alignment also influence the variability of the normalised data. Rather than introducing uncertainty, in this case the uncertainty is redistributed depending on the relative magnitude of the measurements (Figures 3A and 3B). In particular, the variability of high intensity measurements is reduced, while the variability of low intensity measurements is increased. This redistribution is due to the fact that normalised data points depend on the data points from other conditions or even from other replicates, as highlighted by the random variable of the normalised data in Equation (5).

A consequence of this redistribution is also that normalisations by sum and by optimal alignment have an impact on statistical testing. By applying a two-tailed t-test we observed an increase in sensitivity and decrease in specificity, when testing conditions with high intensity measurements (Figures 5A,B). Because more false positives are detected, the distinction of differences between two data points with high intensity measurements becomes less reliable than before normalisation. The alterations of sensitivity and specificity are inverted when data points with low intensity measurements are tested. These results imply that when these normalisations are applied, it is necessary to pay attention to whether high intensity or low intensity data points are tested and interpret the results accordingly. It is also possible to envision the definition of a data transformation or modified t-test to tune sensitivity and specificity based on the relative magnitude of the measurements tested, and calibrate the number of false positives to 5% of the cases.

Our findings also have implications for the use of Western blot data for mathematical model training and validation. In this setting, data is compared to the output of a model and appropriate values for the parameters of the model are identified, aiming to obtain the best possible agreement between data and output [4]. Because data normalisation has an influence on the distribution of the normalised data, we advise to normalise also the model output before comparing it to the data, when the nature of the mathematical model permits it. This should allow for a fair comparison between output and data, because in principle they would be subject to the same data transformation.

Although the quantitative use of Western blotting is now widespread, published articles often lack the details of how Western blot results were quantified and how biological replicates were compared to obtain statistics [23]. We hope that the results in this paper will serve as a reference and encourage scientists to include in future publications what we demonstrate to be critical information. To this end, based on our results, we wrote a short manual of one page that contains a step-by-step guide to help biologists choose the normalisation strategy that is the most appropriate to their case.

SUPPORTING INFORMATION

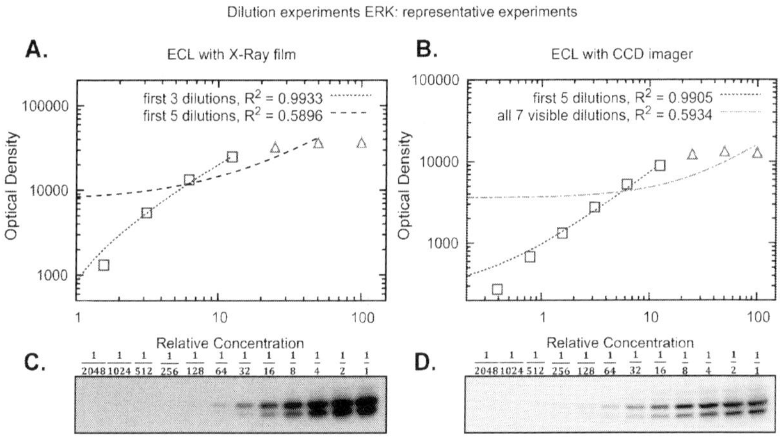

Figure S1. Signal linearity of ERK obtained by different Western blot detection systems. Shown are representative results from three independent experiments of Western blots containing 2-fold serial dilution of cell lysate. ERK was detected by (A,C) ECL with X-ray film and (B,D) ECL with CCD imager. Blue squares indicate data points that are linear, while red triangles indicate data points outside the linear range of detection. To highlight linear and non-linear data we use linear trend lines, reporting the coefficient of determination R^2. In (A,B) data are in log-log scale to improve visualisation.

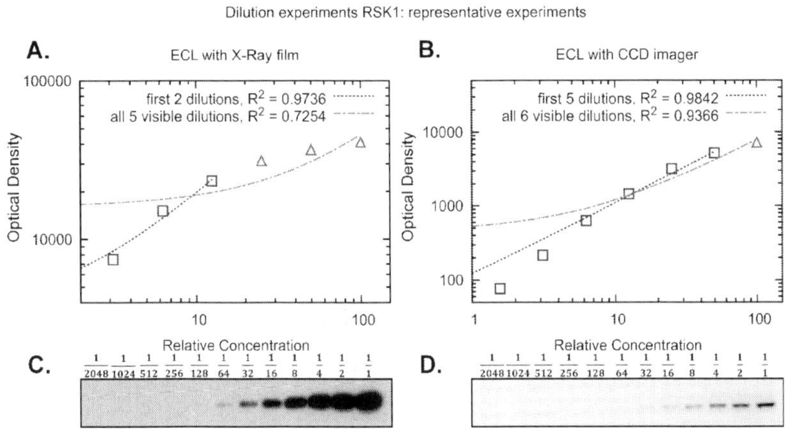

Figure S2. Signal linearity of RSK1 obtained by different Western blot detection systems. Shown are representative results from three independent experiments of Western blots containing 2-fold serial dilution of cell lysate. RSK1 was detected by (A,C) ECL with X-ray film and (B,D) ECL with CCD imager. Blue squares indicate data points that are linear, while red triangles indicate data points outside the linear range of detection. To highlight linear and nonlinear data we use linear trend lines, reporting the coefficient of determination R^2. In (A,B) data are in log-log scale to improve visualisation.

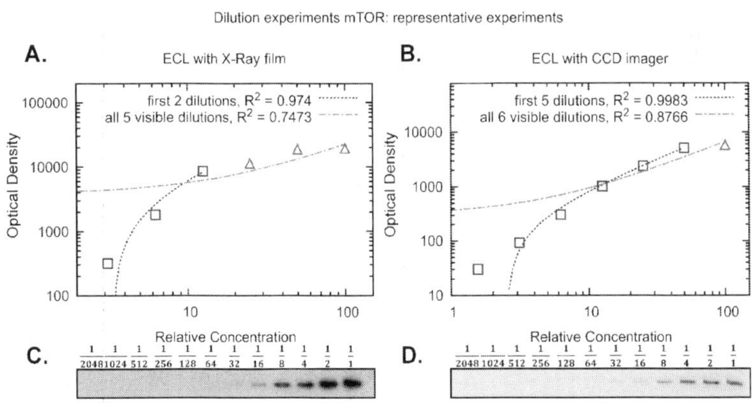

Figure S3. Signal linearity of mTOR1 obtained by different Western blot detection systems. Shown are representative results from three independent experiments of Western blots containing 2-fold serial dilution of cell lysate. Protein mTOR1 was detected by (A,C) ECL with X-ray film and (B,D) ECL with CCD imager. Blue squares indicate data points that are linear, while red triangles indicate data points outside the linear range of detection. To highlight linear and non-linear data we use linear trend lines, reporting the coefficient of determination R^2. In (A,B) data are in log-log scale to improve visualisation.

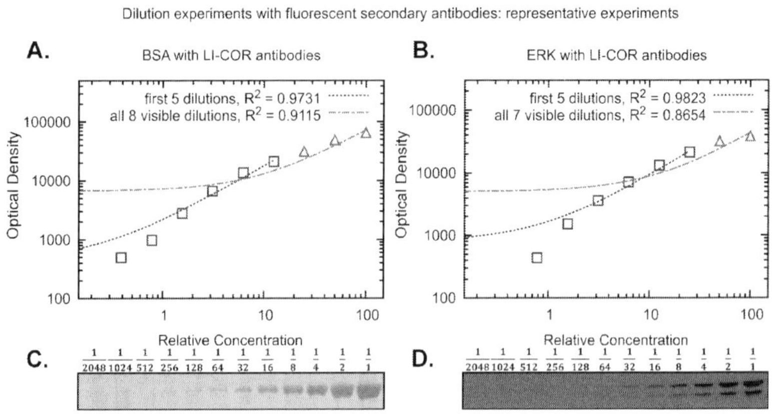

Figure S4. Signal linearity of BSA and ERK obtained by fluorescent secondary antibodies. Shown are representative results from three independent experiments of Western blots containing 2-fold serial dilution of (A,C) BSA and (B,D) cell lysate. BSA and ERK were detected using fluorescent secondary antibodies. Blue squares indicate data points that are linear, while red triangles indicate data points outside the linear range of detection. To highlight linear and non-linear data we use linear trend lines, reporting the coefficient of determination R^2. In (A,B) data are in log-log scale to improve visualisation.

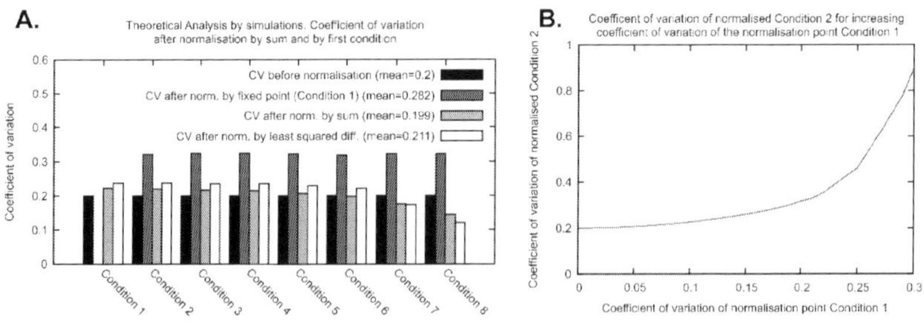

Figure S5. Effect of the normalisation on the coefficient of variation of the normalised data. (A) CVs are shown for the distribution of the simulated data before normalisation, after normalisation by first condition, after normalisation by sum of all data points in a replicate and after normalisation by least squared differences. The mean coefficient of variation is computed as the average across the eight conditions. Mean and standard deviation of the data before normalisation is given in Figure 3A of the main text, and here is normally distributed. (B) Before normalisation, the response to Condition 2 has a coefficient of variation of 0.2, as shown in Figure 3A of the main text. Condition 2 is then normalised by fixed point, with Condition 1 as normalisation point. Here we show how the coefficient of variation of normalised Condition 2 changes for increasing coefficient of variation of the normalisation point Condition 1.

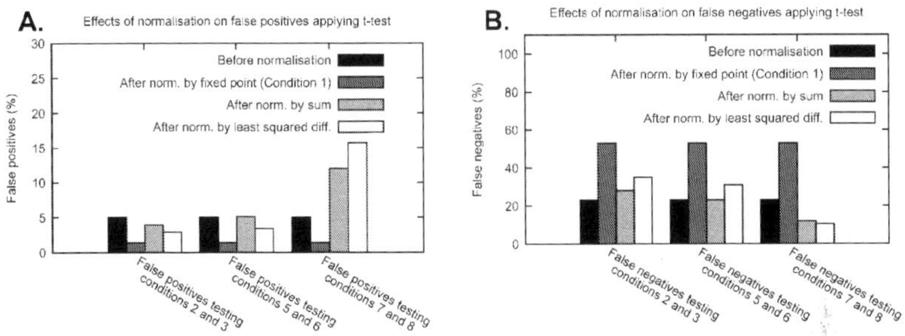

Figure S6. Effects of normalisation on false positives and false negatives when applying t-test for equality of the mean. (A) We consider responses to eight conditions with normal distributions with CV of 0.2 and means of the conditions from 1 to 8 equal to: 1, 2, 2, 4, 7, 7, 18, 18. A number n=5 of sampled replicates are obtained from these distributions and normalised using the normalisations above. Using these replicates before and after normalisation, conditions are tested using a two-tailed t-test with threshold p-value of 0.05. We repeat this procedure a large number of times and estimate the percentage of false positives. (B) In analogy with (A), we estimate the number of false negatives considering means of the conditions from 1 to 8 equal to: 1, 2, 3, 4, 7, 10.5, 18, 27. Notice that for a fair comparison, when testing two conditions, one has a mean that is always 2/3 the mean of the other, e.g. Condition 5 has mean 7 and Condition 6 has mean 10.5, with 7/10.5=2/3.

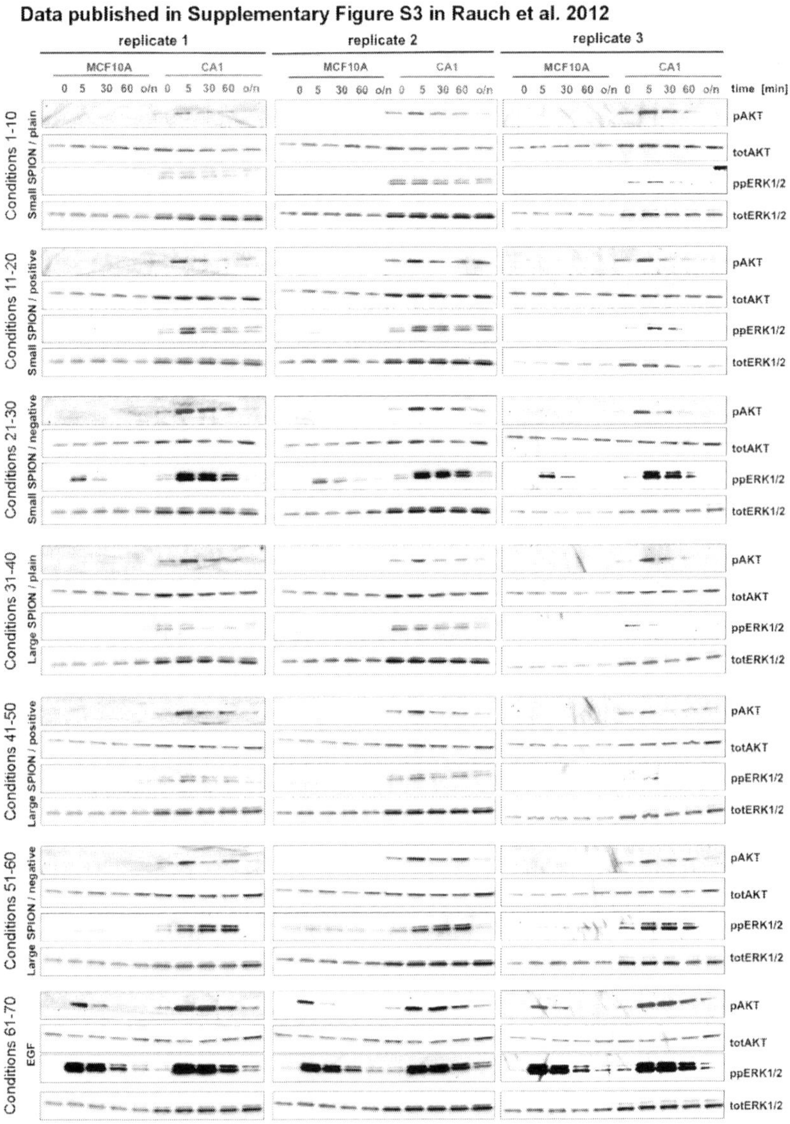

Figure S7. Figure S3 of [25] . Experimental data used in Figures 3C and 4. The experiments shown in Figure S5 were performed as described by Rauch et al. in [25].

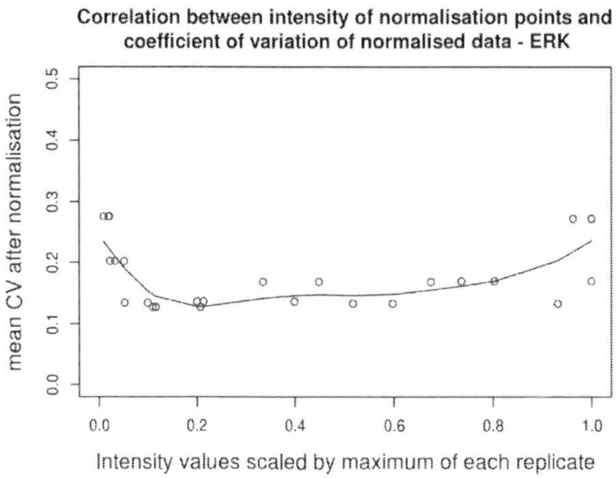

Figure S8. Correlation between the intensity of the normalisation points and the CV of the normalised data. Using data from the three replicates of the ERK dilution experiments detected with CCD imager, we tested every point on a blot as normalisation point. For each resulting normalisation we computed the average of the CV of the normalised data points, and plotted the value of each data point (scaled so that the maximum of each replicate is equal to 1) against the average CV obtained by normalising with the corresponding data point. The result shows how the intensities of each normalisation point chosen correlate with the variability of the normalised data.

ACKNOWLEDGMENTS

We would like to thank Dr. Maria Luisa Guerriero for the helpful discussion.

FUNDING STATEMENT

This work was supported by the Science Foundation Ireland under Grant No. 06/CE/B1129. The funders had no role in study design, data collection and analysis, decision to publish, or preparation of the manuscript.

REFERENCES

1. Burnette WN (1981) "Western blotting". : electrophoretic transfer of proteins from sodium dodecyl sulfate—polyacrylamide gels to unmodified nitrocellulose and radiographic detection with antibody and radioiodinated protein A. Anal Biochem 112: 195–203.
2. Towbin H, Staehelin T, Gordon J (1979) Electrophoretic transfer of proteins from polyacrylamide gels to nitrocellulose sheets: procedure and some applications. Proc Natl Acad Sci U S A 76: 4350–4354.
3. Kitano H (2002) Computational systems biology. Nature 420: 206–210.
4. Maiwald T, Timmer J (2008) Dynamical modeling and multi-experiment fitting with PottersWheel. Bioinformatics 24: 2037–2043.
5. Zar J (1998) Biostatistical Analysis (4th Edition): {Prentice Hall}.
6. Wang YV, Wade M, Wong E, Li YC, Rodewald LW, et al. (2007) Quantitative analyses reveal the importance of regulated Hdmx degradation for p53 activation. Proc Natl Acad Sci U S A 104: 12365–12370.
7. Heidebrecht F, Heidebrecht A, Schulz I, Behrens SE, Bader A (2009) Improved semiquantitative Western blot technique with increased quantification range. Journal of Immunological Methods 345: 40–48.
8. Speed TP (2003) Statistical analysis of gene expression microarray data. Boca Raton FL: Chapman & Hall/CRC xiii, 222 p., 224 p. of plates p.
9. Bolstad BM, Irizarry RA, Astrand M, Speed TP (2003) A comparison of normalization methods for high density oligonucleotide array data based on variance and bias. Bioinformatics 19: 185–193.
10. Li C, Wong WH (2001) Model-based analysis of oligonucleotide arrays: expression index computation and outlier detection. Proc Natl Acad Sci U S A 98: 31–36.
11. Zhang R, Yang D, Zhou C, Cheng K, Liu Z, et al. (2012) beta-actin as a loading control for plasma-based Western blot analysis of major depressive disorder patients. Anal Biochem 427: 116–120.
12. Liu NK, Xu XM (2006) beta-tubulin is a more suitable internal control than beta-actin in western blot analysis of spinal cord tissues after traumatic injury. J Neurotrauma 23: 1794–1801.

13. Wu Y, Wu M, He G, Zhang X, Li W, et al. (2012) Glyceraldehyde-3-phosphate dehydrogenase: a universal internal control for Western blots in prokaryotic and eukaryotic cells. Anal Biochem 423: 15–22.
14. Bruning U, Fitzpatrick SF, Frank T, Birtwistle M, Taylor CT, et al. (2012) NFkappaB and HIF display synergistic behaviour during hypoxic inflammation. Cell Mol Life Sci 69: 1319–1329.
15. Lange F, Rateitschak K, Fitzner B, Pohland R, Wolkenhauer O, et al. (2011) Studies on mechanisms of interferon-gamma action in pancreatic cancer using a data-driven and model-based approach. Mol Cancer 10: 13.
16. Witt J, Konrath F, Sawodny O, Ederer M, Kulms D, et al. (2012) Analysing the role of UVB-induced translational inhibition and PP2Ac deactivation in NF-kappaB signalling using a minimal mathematical model. PLoS One 7: e40274.
17. Neumann L, Pforr C, Beaudouin J, Pappa A, Fricker N, et al... (2010) Dynamics within the CD95 death-inducing signaling complex decide life and death of cells. Mol Syst Biol 6.
18. Wang CC, Cirit M, Haugh JM (2009) PI3K-dependent cross-talk interactions converge with Ras as quantifiable inputs integrated by Erk. Mol Syst Biol 5: 246.
19. Schilling M, Maiwald T, Bohl S, Kollmann M, Kreutz C, et al. (2005) Computational processing and error reduction strategies for standardized quantitative data in biological networks. FEBS J 272: 6400–6411.
20. Aksamitiene E, Hoek JB, Kholodenko B, Kiyatkin A (2007) Multistrip Western blotting to increase quantitative data output. Electrophoresis 28: 3163–3173.
21. Kreutz C, Bartolome Rodriguez MM, Maiwald T, Seidl M, Blum HE, et al. (2007) An error model for protein quantification. Bioinformatics 23: 2747–2753.
22. Rocke DM, Durbin B (2001) A model for measurement error for gene expression arrays. J Comput Biol 8: 557–569.
23. Gassmann M, Grenacher B, Rohde B, Vogel J (2009) Quantifying Western blots: pitfalls of densitometry. Electrophoresis 30: 1845–1855.
24. Box GEP, Muller ME (1958) A Note on the Generation of Random Normal Deviates. Annals of Mathematical Statistics 29: 610–611.

25. Rauch J, Kolch W, Mahmoudi M (2012) Cell Type-Specific Activation of AKT and ERK Signaling Pathways by Small Negatively-Charged Magnetic Nanoparticles. Sci Rep 2.
26. Bartlett MS (1935) The Effect of Non-Normality on the t Distribution. Mathematical Proceedings of the Cambridge Philosophical Society 31: 223–231.
27. Boneau CA (1960) The effects of violations of assumptions underlying the test. Psychol Bull 57: 49–64.

8

The Design of a Quantitative Western Blot Experiment

Sean C. Taylor[1] and Anton Posch[2]

[1]Bio-Rad Laboratories, 1329 Meyerside Drive, Mississauga, ON, Canada
[2]Bio-Rad Laboratories GmbH, Heidemann Street 164, 80939 Munich, Germany

ABSTRACT

Western blotting is a technique that has been in practice for more than three decades that began as a means of detecting a protein target in a complex sample. Although there have been significant advances in both the imaging and reagent technologies to improve sensitivity, dynamic range of detection, and the applicability of multiplexed target detection, the basic technique has remained essentially unchanged. In the past, western blotting was used simply to detect a specific target protein in a complex mixture, but now journal editors and reviewers are requesting the quantitative interpretation of western blot data in terms of fold changes in protein expression between samples. The calculations are based on the differential densitometry of the

Citation: The Design of a Quantitative Western Blot Experiment. Sean C. Taylor and Anton Posch. Hindawi Publishing Corporation BioMed Research International Volume 2014, Article ID 361590, 8 pages http://dx.doi.org/10.1155/2014/361590

associated chemiluminescent and/or fluorescent signals from the blots and this now requires a fundamental shift in the experimental methodology, acquisition, and interpretation of the data. We have recently published an updated approach to produce quantitative densitometric data from western blots (Taylor et al., 2013) and here we summarize the complete western blot workflow with a focus on sample preparation and data analysis for quantitative western blotting.

INTRODUCTION

Proteomic technologies such as two-dimensional electrophoresis and mass spectrometry are valuable tools in semiquantitative protein profiling studies in order to identify broad expression patterns enabling a better understanding of molecular events, signaling pathways and mechanisms [1]. The resulting data are typically confirmed by a second, independent method such as western blotting. Western blotting was introduced by Towbin et al. [2] in 1979 and has since become a common technique used in research laboratories globally for the immunodetection and quantitation of specific proteins in complex cell homogenates. Over the past three decades, the sensitivity, robustness, and flexibility of the corresponding indicator systems have increased significantly [3, 4]. In addition, the ongoing development of detection media and reagents has provided the scientific community with ultrasensitive imaging systems giving broad dynamic range of detection enabling precise and accurate quantitation of signals from both low and high expressing proteins from the same blot. Although labs have been quick to purchase the latest detection technologies and reagents for western blotting, the associated techniques used to produce the densitometric data have not evolved leading to published data that are difficult or impossible to interpret or reproduce [5–7].

In order to obtain quantitative data from western blots, a rigorous methodology must be used as previously described [8]. Briefly, the validation of antibodies (Ab) is critical both to assure that the Ab/antigen interaction is specific and correct and to determine the dilution factor of samples that is required for protein loading in the quantitative linear dynamic range for each antibody. Furthermore, the appropriate selection of normalization method (based on reference signals obtained either by housekeeping proteins (HKPs) after immunochemical staining or total protein (TP) intensity on blotting membranes after total protein staining) must be considered to assure that the reported fold changes of the target protein are not an artifact of reference signal. Thus, data normalization is crucial to identify and correct experimental errors where reference instability becomes increasingly important with the measurement of smaller differences in target protein expression between samples [9]. The direct effect of poor normalization is evident when sample loading above 10 μg of a total protein lysate per lane is required because traditional loading HKPs such as GAPDH, actin, and tubulin are grossly overloaded and therefore not serving the purpose of data normalization [8, 9]. Also, these HKPs can be affected by the treatment conditions of the experiment giving skewed results for target protein expression that do not reflect the biology of the tested samples [10–15]. Alternatively, normalization by total, blot-transferred protein has recently been shown to give excellent data for typical total protein lysates [16].

Here, we describe some general techniques to produce good quality protein samples with minimal degradation for improved reproducibility between experiments. Also summarized are the basic steps of quantitative western blotting and a standardized approach to calculating the associated densitometric data from multiple blots.

CAREFUL EXPERIMENTAL DESIGN PRODUCES RELIABLE AND REPRODUCIBLE DATA

Unlike DNA-based assays that measure a predictable type of molecule that is typically stable in a variety of conditions, proteins can vary significantly in their expression, stability, conformation, and activity under different buffer and experimental conditions. Furthermore, the presence of contaminating proteins in a homogenate can greatly affect the integrity and activity of target proteins [17]. Care must therefore be taken in the design of any protein-based assay to ensure that the apparent differences between case and control samples are not an artifact of the experimental conditions or sample handling. Factors that can have a major influence on the proteome include incubation time and temperature, as well as the parameters for processing samples such as the amount of time between tissue collection and subsequent freezing or even the conditions and timing for thawing tissue or cell pellets (Table 1).

Table 1. Major design parameters for protein-based assays. A solid experimental design that maps the procedure, controls, replicates, experimental conditions, and sampling handling guidelines sets the foundation for production of solid, quantifiable data.

Experiment procedure	Control groups	Replicates	Experiment conditions	Sample handling
Disease or treatment groups	Time course study (i.e.,)	Biological (difft sample per well)	Growth conditions (media and time or OD)	Precise time to harvest cells or tissues
Target proteins implicated	Normal versus disease (i.e., normal)	Technical (same sample per well)	Days of embryonic development	Sample extraction method
Potential internal controls	Untreated versus drug treated (i.e., untreated)		Amount per mass of drug or compound	Preservation method and time
These first three steps define the following: experimental parameters and the goals and the samples based on literature or experimental data typically from broad microarray or proteomics experiments			Sex, phenotype Incubation time	Thaw and homogenization procedure Total protein extraction procedure

SAMPLE PREPARATION

There are several pitfalls associated with sample preparation that can directly affect the density of bands on a western blot including:

- improper handling of tissue or cell specimens resulting in variable degradation and/or expression of proteins between samples,
- inadequate detergents, salts, and protease inhibitors in the lysis buffer,
- poor homogenization technique.

Since protein lysates are highly complex with contaminants such as cellular or tissue debris, fats, hydrophobic protein aggregates, nucleic acids, and proteases that can directly and negatively affect the results from western blots, it is important to use cell lysis buffers and homogenization techniques that eliminate their effects [17]. In general, homogenization buffers containing nonionic detergents such as NP-40 and Triton X-100 are less harsh than ionic detergents, such as SDS and sodium deoxycholate. Salts such as NaCl or KCl are typically added to a concentration of 100 to 150 mM to prevent protein aggregation. RIPA buffer (1% NP-40 or Triton X-100, 1% sodium deoxycholate, 0.1% SDS, 150 mM NaCl, 50 mM Tris-HCl, pH 7.8, 1 mM EDTA) and complete mini protease inhibitor cocktail tablets (Roche Applied Science) in combination with either mechanical or manual homogenization instruments have been used to produce homogenates that give solid data for western blot assays.

The proper choice of tissue homogenization technique is a prerequisite for a successful western blot assay and the method employed entirely depends on the sample type (i.e., brain versus muscle versus liver tissue as opposed to plated or suspended cells) [18, 19]. A good example of a tissue lysis protocol is as follows:

- Snap-freeze the tissue in liquid nitrogen and then dice tissue into 1 mm pieces with a scalpel in a mortar on dry ice. Ensure that the scalpel or grinder is also frozen on dry ice to keep the cut or ground tissue close to the temperature of dry ice throughout the procedure.
- Add the diced/ground tissue to ice-cold RIPA buffer.
- Transfer the tissue preparation to an ice-cold Dounce tissue homogenizer (Wheaton) and Dounce 25x on ice.
- Sonicate (Tekmar Sonic Disrupter) the Dounce-homogenized tissue on ice for 5 × 20 seconds at 50% power and clear the extracts by centrifugation at 34,000 ×g at 4°C for 30 minutes.
- Transfer the supernatant to a new tube and perform protein assay (see below).
- Store the supernatants at −80°C or in liquid nitrogen for long term storage.

For cell lysis, add the pelleted cells (in the case of cell suspensions) to ice-cold RIPA buffer or for plated cells, add the ice-cold RIPA buffer directly to the plate after washing the cells, and scrape and pipette the cells up and down. Continue with step (3) above.

The total protein concentration of the homogenate from either cell or tissue lysates should be measured using a detergent compatible protein assay such as the RC DC protein assay from Bio-Rad. Ideally, the homogenates would be diluted to a concentration of at least 2 mg/mL which would permit loading between 10 μg and 80 μg per lane of a 1 mm thick mini polyacrylamide SDS-gel.

DETERMINE THE LINEAR DYNAMIC RANGE OF PROTEIN LOADING

Most labs load a random amount of protein in each lane of a gel for western blotting that is typically between 10 μg and 100 μg of total lysate and there is typically no scientific basis for choosing this amount. This often results in the overloading of highly expressed, target proteins and particularly the loading controls that are used to normalize the data. This typically gives uniform band densities between lanes for the housekeeping proteins which is not due to consistent protein loading but rather from overloading the membrane with the target protein (Figure 1). To alleviate the effect of membrane saturation, a standard curve of protein load versus band density should be produced for each target protein. This can be accomplished by making a 1/2 dilution series of a pooled sample from all the lysates in the study group starting from 100 μg protein load over at least 12 dilutions on a TGX stain-free SDS-gel (Bio-Rad). Stain-free detection on the ChemiDoc MP (Bio-Rad) camera system can be used to verify the loading, quality, and separation of the homogenate followed by transfer to a low fluorescent PVDF membrane using the Trans Blot Turbo (Bio-Rad) protein transfer system [8].

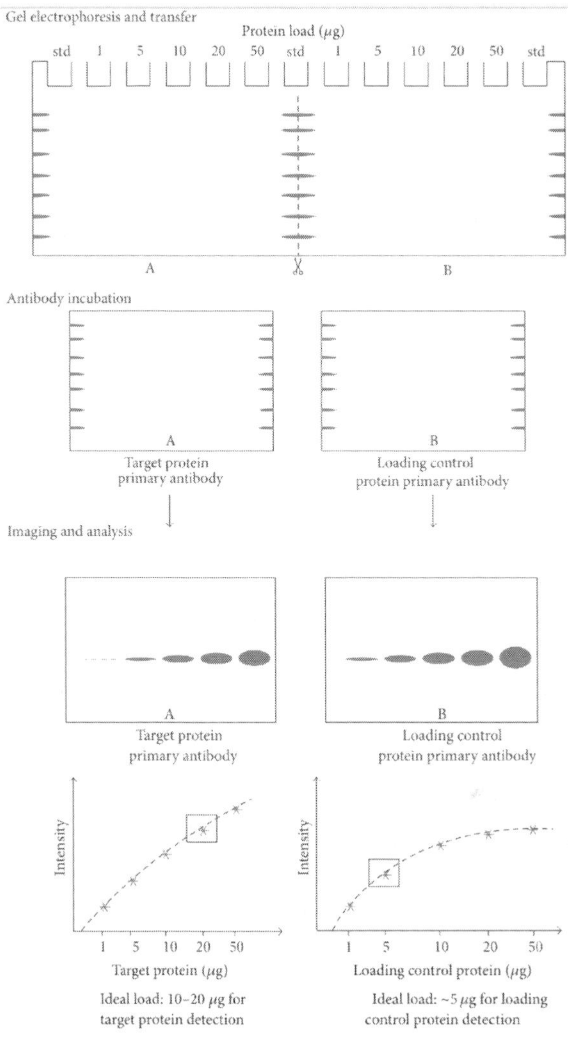

Figure 1. Reliable western blot data can only be generated when the proper sample amount of protein is used. Loading too much protein leads to signal saturation in western blots, yet too little produces weak signals. Once the experimental setup and conditions are established for the assay, do not change the sample load, transfer method, transfer time, antibody dilution, antibody incubation time, or temperature in subsequent experiments as these factors may significantly change the detection signals.

A typical methodology for determination of the appropriate loading for protein samples follows:

- Transfer and blot accordingly by incubating the target protein primary antibody and associated secondary to each blot with at least four, 3-minute wash steps between each incubation.

- Add an imager-compatible, chemiluminescence substrate such as Clarity (Bio-Rad) to develop the immunochemical signal and capture the signal using a CCD-camera-based imager such as the ChemiDoc MP (Bio-Rad).
- Image the blots using software that provides accurate, background-subtracted densitometric tools such as Image Lab (Bio-Rad) and produce a plot of relative density versus fold dilution for each primary antibody.
- Validate the antibodies by determining their linear dynamic range (i.e., the range in which a consistent, 1/2 decrease in density is obtained).
- Select the protein load for each antibody that corresponds to the middle of the linear dynamic range.

Dilution of the individual samples in the study group to the middle point in the linear dynamic range of the pooled sample for each antibody may mean that the individual protein samples require widely different dilutions for each antibody. This will assure that the densitometric data for each target protein will be within the linear dynamic (quantitative) range to give accurate and reproducible results reflecting the true biology between samples in the study set (Figure 2). Inappropriate loading of samples may result in no quantifiable difference between the samples for a given target simply due to overloading the membrane.

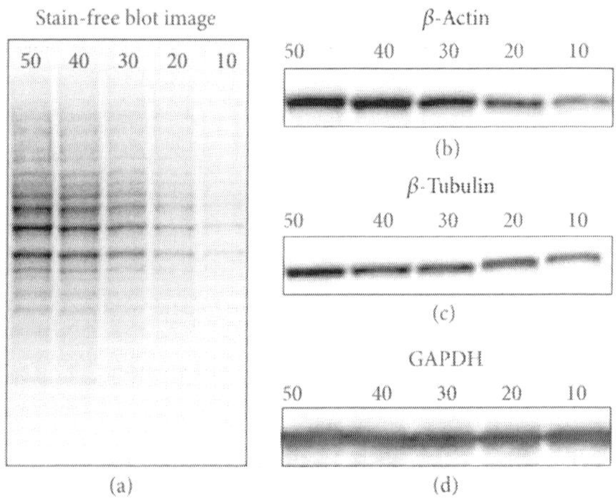

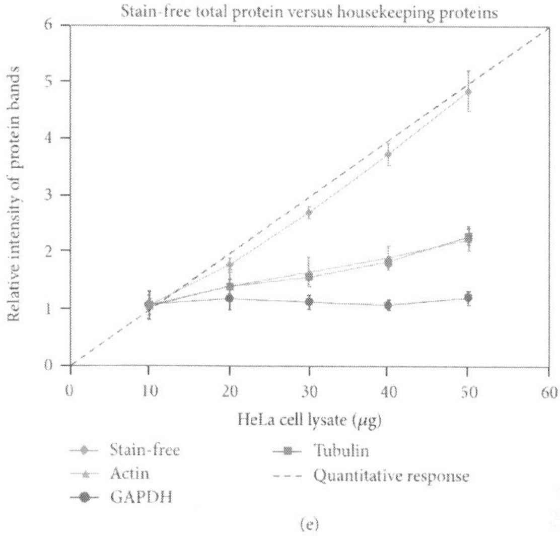

Figure 2. (From [9] with permission from the authors and Bio-Rad.) Linearity comparison of stain-free total protein measurement and immunodetection of three housekeeping proteins in 10–50 μg of HeLa cell lysate. On the left are representative images of (a), stain-free blot and the chemi blots for (b), β-actin; (c), β-tubulin and (d), GAPDH. Lane labels correspond to total protein load (μg). Although the actin and tubulin signals appear linear, the densitometric ratio was far below the predicted "quantitative response" of actual loading whereas the stain-free signal correlated to the expected result (e).

DETERMINE THE APPROPRIATE REFERENCE SIGNAL FOR DATA NORMALIZATION

A good reference signal or "loading control" is one that is coexpressed with the target protein within the same sample and consistently expressed between samples. HKPs such as tubulin, β-actin, and GAPDH have traditionally served as loading controls, but there are three potential drawbacks to using such controls.

- HKPs may not be expressed in a uniform manner between the experimental conditions which will give erroneous results [10–15]. The same issue has been found with reference genes used for qPCR where the selection of unstable targets has led to opposite results when contrasted with stable targets [20, 21].

- HKPs are highly abundant in lysates and have typically saturated the membrane for samples loaded in excess of about 4 μg per lane (see previous section) (Figure 2). This would give these proteins the "appearance" of good normalization controls because the densities of the associated bands would all be similar between lanes as an "artifact" of membrane saturation.
- Data normalization with HKPs relies only on one data point and provides a poor reflection of possible process inconsistencies.

Given the problems that arise with HKP controls, alternative methods for normalization have been sought out by the scientific community and we propose that an excellent loading control (LC) should meet the following criteria.

- (1)It has good responsiveness (1 : 1) to changes in total protein amount of individual samples.
- It is insensitive to the influence of various physiological conditions and treatments and therefore must be quantified from the membrane itself to take into account the effect of transfer efficiency [22, 23].
- Acquisition of the LC would ideally be possible at all phases of the western blot process (i.e., visualization of protein on the pretransfer gel lanes, posttransfer blot, and posttransfer gel lanes) thus enabling a consistent process control.
- Acquisition of the LC must be fast, easy, and highly reproducible. (
- No lengthy process should be required for the optimization and establishment of LC.
- The method for LC detection should be compatible with immunochemical staining.

To address these issues, the scientific community is now adopting the use of total lane density from the blot-transferred protein as a means of data normalization [24–26]. There are a number of stains that can be used to visualize, image, and quantify the transferred protein on the blot including Ponceau S, Coomassie R-350, Amido Black, MemCode, and Deep Purple. However, each of these stains has individual issues of being poorly reproducible on a day-to-day basis, limited dynamic range, and restricted compatibility with blotting membranes and immunochemical staining [25].

More recently the Stain-Free technology (Bio-Rad) has been introduced [24–26] and meets all six of the criteria mentioned above for a good LC with a linear dynamic range between about 10 and 80 μg of total protein load from a typical cell or tissue lysate [8, 9]. This permits the use of total lane density from the stain-free blot for normalization between lanes for most western blot studies.

The technique for total lane normalization using the stain-free assay technology has been well-described but briefly it is as follows:

- The quality of the electrophoretic separation can be verified within a couple of minutes. After UV-activation, the protein bands are visible in the gel and can be recorded with a camera system. The generated fluorescent signal remains stable over a couple of hours.
- The blot is imaged immediately after transfer to verify the transfer of protein from each lane.
- The image data from the total density of all the blot-transferred protein bands per lane is then recorded using Image Lab software by selecting a single band in each lane and stretching the band width to cover all the volume peaks in the lane profile.
- The background rolling disc is adjusted to a low value (between 1 and 5) for all the lanes to assure that only the total background subtracted density from the sum of all the bands per lane is acquired for normalization.

In addition, Stain-Free technology is compatible with both nitrocellulose and PVDF membranes and data normalization with SF blot images is based on many data points which is superior to HKPs.

DATA ANALYSIS: THE BACKGROUND SUBTRACTION PROBLEM

Background subtraction is a common reason to obtain variable or incorrect data from western blots [27]. Using traditional densitometric analysis methods such as volume analysis from boxed bands and background, variations in background-subtracted data can arise since the background is not subtracted from the same box in which the band resides. Furthermore, the box in which a band is selected always contains density from both the band and the associated background which becomes more prevalent with low density bands [8]. The combination of these factors can result in highly variable data when testing samples with a differentially expressed target protein using a nonspecific antibody with high background. A good alternative to volume analysis using boxes is a rolling disc background subtraction algorithm coupled with a lane profile tool (Figure 3). Image Lab software is designed with both of these tools that can be used simultaneously to ensure that the appropriate band width and lane background is selected for each lane (Figure 3).

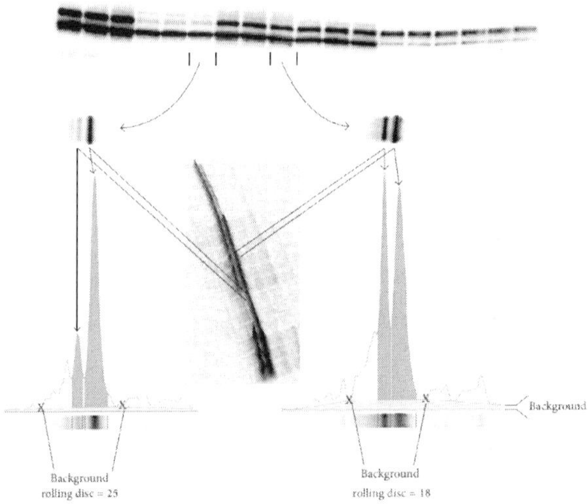

Figure 3. (From [8] with permission from the authors and Bio-Rad.) Image acquisition and densitometric analysis. Image Lab software version 5.0 (Bio-Rad) was used for image acquisition and densitometric analysis of the gels, blots, and film in this study. The software interprets the raw data in three dimensions with the length and width of the band defined by the "Lanes and Bands" tool in concert with the "Lane Profile" tool such that the chemiluminescent signal emitted from the blot is registered in the third dimension as a peak rising out of the blot surface. The density of a given band was measured as the total volume under the three-dimensional peak, which could be viewed in two dimensions using the "Lane Profile" tool to adjust the precise width of the band to account for the area under the shaded peak of interest. Background subtraction was set by using the rolling disc setting in the "Lanes" tool. The rolling disc values were set such that the background was subtracted under the band (i.e., peak) of interest in a uniform manner between the lanes of a given blot. In this case, the rolling disc for the two lanes analyzed was set to 18 and 25, respectively, such that the peaks of interest were cut at a consistent level between the markers shown with an "X".

COMPUTATIONAL ANALYSIS OF DENSITOMETRIC DATA

There are numerous calculations from densitometric data using formulas buried in EXCEL spreadsheets spanning multiple worksheets and files to obtain quantitative data from western blots. It is often difficult to follow the basis of the calculations and we have found that when lab members are faced with the direct question of how to work up the raw data obtained from western blots to publishable results, there is often confusion. The analysis of western blot data can be accomplished using a very similar methodology to qPCR by calculating relative, normalized protein expression as described in the following steps (Tables 2 and 3).

- For each blot, multiply the background subtracted density (volume in Image Lab software) of the target protein (TP) in each lane by the ratio of density of the loading control (LC) (either housekeeping protein or total lane density) from a control sample loaded into lane 1 of all the study blots to the other lanes in the gel. This will give the normalized density to the loading control (NDL) (Table 2). The control sample is typically a pooled homogenate from all of the samples in a given study aliquoted into multiple tubes to permit the loading of a fresh control sample in lane 1 of each study blot.
- Calculate the fold difference (FD) for each biological/technical replicate by dividing NDL from each lane by the NDL from the control sample in lane 1 (Table 2).
- Determine the average FD and associated P values for the biological replicates by importing the FD from step (2) above into a statistical analysis software package such a PRISM or Analyze IT (Table 3).

Table 2. Computational analysis of densitometric data (*density data from a control sample loaded in the first lane of each gel/blot for a given study).

Biological replicates	Sample	Density (volume)			
		Target protein (TP)	Loading control (LC)	Normalized density to LC (NDL) (TP* LC/LC*)	Fold difference (NDL/NDL*)
M1	$T = 0$	8	4*	(8 × 4/4) = 8.0*	(8/8) = 1.0
M1	$T = 1$	45	5	(45 × 4/5) = 36.0	(36/8) = 4.5
M1	$T = 2$	90	6	(90 × 4/6) = 60.0	(60/8) = 7.5
M2	$T = 0$	6	5	(6 × 4/5) = 4.8	(4.8/8) = 0.6
M2	$T = 1$	40	5	(40 × 4/5) = 32.0	(32/8) = 4.0
M2	$T = 2$	88	7	(88 × 4/7) = 50.3	(50.3/8) = 6.3
M3	$T = 0$	10	4	(10 × 4/4) = 10.0	(10/8) = 1.3
M3	$T = 1$	48	6	(48 × 4/6) = 32.0	(32/8) = 4.0
M3	$T = 2$	92	6	(92 × 4/6) = 61.3	(61.3/8) = 7.7

Table 3. Statistical analysis of densitometric data.

FD by time	n	Mean	SE	Pooled SE	SD
$T = 0$	3	0.95	0.189	0.290	0.33
$T = 1$	3	4.17	0.167	0.290	0.29
$T = 2$	3	7.15	0.435	0.290	0.75
Source of variation	Sum squares	DF	Mean square	F statistic	P
Time	57.70	2	28.85	114.03	<0.0001
Residual	1.52	6	0.25		
Total	**59.22**	**8**			
Tukey contrast	Difference	95% CI			
$T = 0$ v $T = 1$	−3.22	−4.48	to −1.96	(Significant)	
$T = 0$ v $T = 2$	−6.20	−7.46	to −4.94	(Significant)	
$T = 1$ v $T = 2$	−2.98	−4.24	to −1.72	(Significant)	

CONCLUSIONS

Accurate densitometric analysis of western blots is achieved by a combination of good sample preparation, technique, detection method, software, and analysis. By following the steps outlined here, the outcome of a given experiment should produce excellent results. For highest data reproducibility and integrity the application of Stain-Free technology is highly recommended because this approach offers a novel and unique quality control tool for data normalization in a standardized manner in western blotting workflows.

ABBREVIATIONS

Ab:	Antibody
CCD:	Charge-coupled device
CL:	Chemiluminescence
FD:	Fold difference
HKPs:	Housekeeping proteins
LC:	Loading control
NDL:	Normalized density to the loading control
qPCR:	Real-time polymerase chain reaction
SF:	Stain-free
TP:	Target protein
WB:	Western blotting.

CONFLICT OF INTERESTS

The authors declare that there is no conflict of interests regarding the publication of this paper.

REFERENCES

1. D. Penque, "Two-dimensional gel electrophoresis and mass spectrometry for biomarker discovery," *Proteomics—Clinical Applications*, vol. 3, no. 2, pp. 155–172, 2009.
2. H. Towbin, T. Staehelin, and J. Gordon, "Electrophoretic transfer of proteins from polyacrylamide gels to nitrocellulose sheets: procedure and some applications," *Proceedings of the National Academy of Sciences of the United States of America*, vol. 76, no. 9, pp. 4350–4354, 1979.
3. A. Dubitsky, D. DeCollibus, and G. A. Ortolano, "Sensitive fluorescent detection of protein on nylon membranes," *Journal of Biochemical and Biophysical Methods*, vol. 51, no. 1, pp. 47–56, 2002.
4. K. McDonald, "Increase Western blot throughput with multiplex fluorescent detection: advice and tips on how to quickly generate reliable and reproducible results," *Pharma*, 2010.
5. J. Bordeaux, A. W. Welsh, S. Agarwal et al., "Antibody validation," *BioTechniques*, vol. 48, no. 3, pp. 197–209, 2010.
6. J. P. Mollica, J. S. Oakhill, G. D. Lamb, and R. M. Murphy, "Are genuine changes in protein expression being overlooked? Reassessing Western blotting," *Analytical Biochemistry*, vol. 386, no. 2, pp. 270–275, 2009.
7. Editorial, "A picture worth a thousand words (of explanation)," *Nature Methods*, vol. 3, no. 4, p. 237, 2006.
8. S. C. Taylor, T. Berkelman, G. Yadav, and M. Hammond, "A defined methodology for reliable quantification of Western blot data," *Molecular Biotechnology*, vol. 55, no. 3, pp. 217–226, 2013.
9. M. Hammond, J. Kohn, K. Oh, P. Piatti, and N. Liu, "A Method for Greater Reliability in Western Blot Loading Controls—Stain-Free Total Protein Quantitation," Bio-Rad Bulletin: 6360, 2013.
10. A. Dittmer and J. Dittmer, " -Actin is not a reliable loading control in Western blot analysis," *Electrophoresis*, vol. 27, no. 14, pp. 2844–2845, 2006.
11. G. M. Aldridge, D. M. Podrebarac, W. T. Greenough, and I. J. Weiler, "The use of total protein stains as loading controls: an alternative to high-abundance single-protein controls in semi-quantitative immunoblotting," *Journal of Neuroscience Methods*, vol. 172, no. 2, pp. 250–254, 2008.

12. M. Rocha-Martins, B. Njaine, and M. S. Silveira, "Avoiding pitfalls of internal controls: validation of reference genes for analysis by qRT-PCR and western blot throughout rat retinal development," *PLoS One*, vol. 7, no. 8, Article ID 43028, 2012.
13. D. Lanoix, J. St-Pierre, A.-A. Lacasse, M. Viau, J. Lafond, and C. Vaillancourt, "Stability of reference proteins in human placenta: general protein stains are the benchmark," *Placenta*, vol. 33, no. 3, pp. 151–156, 2012.
14. S. Greer, R. Honeywell, M. Geletu, R. Arulanandam, and L. Raptis, "Housekeeping genes; expression levels may change with density of cultured cells," *Journal of Immunological Methods*, vol. 355, no. 1-2, pp. 76–79, 2010.
15. R. E. Ferguson, H. P. Carroll, A. Harris et al., "Housekeeping proteins: a preliminary study illustrating some limitations as useful references in protein expression studies," *Proteomics*, vol. 5, no. 2, pp. 566–571, 2005.
16. J. E. Gilda and A. V. Gomes, "Stain-free total protein staining is a superior loading control to -actin for Western blots," *Analytical Biochemistry*, vol. 440, no. 2, pp. 186–188, 2013.
17. Structural Genomics Consortium, China Structural Genomics Consortium, Northeast Structural Genomics Consortium et al., "Protein production and purification," *Nature Methods*, vol. 5, no. 2, pp. 135–146, 2008.
18. L. De Cecco, V. Musella, S. Veneroni et al., "Impact of biospecimens handling on biomarker research in breast cancer," *BMC Cancer*, vol. 9, article 409, 2009.
19. R. J. Simpson, "Homogenization of mammalian tissue," *Cold Spring Harbor Protocols*, 2010.
20. D. Lanoix, A.-A. Lacasse, J. St-Pierre et al., "Quantitative PCR pitfalls: the case of the human placenta," *Molecular Biotechnology*, vol. 52, no. 3, pp. 234–243, 2012.
21. S. Taylor, M. Buchanan, and M. Basik, "A MIQE Case Study—Effect of RNA Sample Quality and Reference Gene Stability on Gene Expression Data," Bio-Rad Bulletin: 6245, 2011.
22. J. Mozdzanowski, P. Hembach, and D. W. Speicher, "High yield electroblotting onto polyvinylidene difluoride membranes from polyacrylamide gels," *Electrophoresis*, vol. 13, no. 1-2, pp. 59–64,

1992.

23. M. W. Bolt and P. A. Mahoney, "High-efficiency blotting of proteins of diverse sizes following sodium dodecyl sulfate-polyacrylamide gel electrophoresis," *Analytical Biochemistry*, vol. 247, no. 2, pp. 185–192, 1997.
24. A. D. Colella, N. Chegenii, M. N. Tea, I. L. Gibbins, K. A. Williams, and T. K. Chataway, "Comparison of stain-free gels with traditional immunoblot loading control methodology," *Analytical Biochemistry*, vol. 430, no. 2, pp. 108–110, 2012.
25. A. Guertler, N. Kunz, M. Gomolka et al., "Stain-free technology as normalization tool in western blot analysis," *Analytical Biochemistry*, vol. 433, no. 2, pp. 105–111, 2013.
26. R. Li and Y. Shen, "An old method facing a new challenge: re-visiting housekeeping proteins as internal reference control for neuroscience research," *Life Sciences*, vol. 92, no. 13, pp. 747–751, 2013.
27. M. Gassmann, B. Grenacher, B. Rohde, and J. Vogel, "Quantifying Western blots: pitfalls of densitometry," *Electrophoresis*, vol. 30, no. 11, pp. 1845–1855, 2009.

9

Characterization of Nora Virus Structural Proteins via Western Blot Analysis

Brad L. Ericson,[1] Darby J. Carlson,[1] and Kimberly A. Carlson[1]

[1]Biology Department, University of Nebraska at Kearney, 2401 11th Avenue, Kearney, NE 68849, USA

ABSTRACT

Nora virus is a single stranded RNA picorna-like virus with four open reading frames (ORFs). The coding potentials of the ORFs are not fully characterized, but ORF3 and ORF4 are believed to encode the capsid proteins (VP3, VP4a, VP4b, and VP4c) comprising the virion. To determine the polypeptide composition of Nora virus virions, polypeptides from purified virus were compared to polypeptides detected in Nora virus infected *Drosophila melanogaster*. Nora virus was purified from infected flies and used to challenge mice for the production of antisera. *ORF3*, *ORF4a*, *ORF4b*, and *ORF4c* were individually cloned and expressed in *E. coli*; resultant recombinant proteins purified and were used to make monospecific antisera. Antisera were evaluated via Western blot against whole virus particles and

Citation: Characterization of Nora Virus Structural Proteins via Western Blot Analysis. Brad L. Ericson, Darby J. Carlson, and Kimberly A. Carlson. Hindawi Publishing Corporation Scientifica Volume 2016, Article ID 9067848, 8 pages http://dx.doi.org/10.1155/2016/9067848

Nora virus infected fly lysates. Viral purification yielded two particle types with densities of ~1.31 g/mL (empty particles) and ~1.33 g/mL (complete virions). Comparison of purified virus polypeptide composition to Nora virus infected *D. melanogaster* lysate showed the number of proteins in infected cell lysates is less than purified virus. Our results suggest the virion is composed of 6 polypeptides, VP3, VP4a, two forms of VP4b, and two forms of VP4c. This polypeptide composition is similar to other small RNA insect viruses.

INTRODUCTION

Nora virus is a picorna-like virus containing a single stranded positive-sense RNA genome ~12,000 bases long. It is related to a few other viruses including an isolate found in the mountain pine beetle, Lotus plant, a parasitoid wasp, and a freshwater mussel [1]. A hallmark of Nora virus infection is the establishment of persistence with little effect on the host [2]. Currently, the virus must be propagated in infected *Drosophila melanogaster* flies as a cell line suitable for its replication is not yet identified. *D. melanogaster* can be infected by a number of viruses. Some are species specific, such as *Drosophila* A, C, and X viruses, Nora virus, and Sigma virus. Other viruses, such as Cricket Paralysis Virus (CrPV), Flock House virus, and Invertebrate Iridescent virus, infect other species of insects, as well as *D. melanogaster* [3]. This makes *Drosophila* a suitable model for virus replication studies.

The Nora virus genome consists of four open reading frames (ORFs). The first three of these overlap each other in alternative reading frames. The fourth open reading frame, based on its position in the genome, appears to be independently read from the other three. Characterization of the virion suggests it is composed of 4 primary proteins, named VP3, and VP4a, VP4b, and VP4c, specified by ORF3 and ORF4 of the viral genome, respectively. The proteins produced from ORF4 are initially translated into a polyprotein and subsequently released by proteolytic processing. In addition, there are several minor protein components of the virion that appear to be derived from these primary proteins [1]. What is not clear is whether these additional polypeptide components are produced by processing events in the infected cell or are artifacts of purification. If they are not artifacts of the purification procedure, then it would be expected that these additional polypeptides would be found in the infected cells of *Drosophila* flies. The objective of this study was to identify the number of polypeptide species that comprise the Nora virus virion by examining the proteins made in infected *D. melanogaster* flies as well as from purified virus utilizing antisera made against whole virus as well as against the individual structural proteins.

Some of the most important aspects of characterizing a new virus are to determine the size, structure, and polypeptide composition of the virion. This involves a number of steps. First, a purification procedure that yields pure virus particles must be established. Second, SDS-PAGE analysis of the purified virus proteins is performed. If the virus is produced in sufficient quantity, Coomassie blue staining is adequate; if not then radioactive labeling of the viral proteins may be needed for detection. Alternatively, antisera can be produced against whole virus particles and the proteins can be detected by Western blot analysis. Antibody reagents have the additional advantage of being useful in virus detection in the infected cell. If the viral genome is sequenced, the predicted viral proteins can individually be cloned, expressed as recombinant proteins, and the recombinant proteins can be used to produce monospecific antisera. Mass spectrometry can also be performed on viral proteins purified from SDS-PAGE gels and compared back to the known nucleotide sequence of the genome to identify viral protein components [1, 4–7].

Characterization of the Nora virus virion shows that two types of particles are readily made in infected *D. melanogaster*. Western blot analyses of proteins from purified virus and from infected cells indicate that the Nora virus virion is composed of 6 proteins, VP3, VP4a, two forms of VP4b, and two forms of VP4c. Additional components of the virion [1] are most likely artifacts of the purification procedure or are produced upon release of the virus particles as final maturation steps.

MATERIALS AND METHODS

Flies and Virus

D. melanogaster strain *witi Rel*^{E23} infected with Nora virus were a kind gift from Dan Hultmark and Jens-Ola Ekström (Umeå University, Umeå, Sweden). Infected flies were reared under standard conditions at 25°C in a 12-hour light, 12-hour diurnal cycle. The identical uninfected strain was reared under identical conditions in a separate incubator and was confirmed as uninfected by reverse transcription polymerase chain reaction [8]. Infected flies were identified by RT-PCR analysis using primers specific for the VP4a gene (5′-ATGCAGAATCCAACACAAACC and 5′-TCCTAACACCGTTCTCTGTTG). Viral titers in flies were estimated using total RNA isolated from flies and RT-qPCR with these same previously validated primers [2] using a SYBR® Green Quantitative RT-qPCR Kit (Sigma) according to the manufacturer's instructions in a 7500 Real Time PCR System (Applied Biosystems).

Virus Purification and Electron Microscopy

Approximately 3,000 Nora virus infected flies were placed in 10 mM Tris, pH 7.4, and 100 mM NaCl (NT buffer) and homogenized on ice with a motor-driven pestle. The resulting lysate was sonicated on ice for four 15-second pulses with 45 seconds between pulses. The debris was removed by two successive centrifugation steps at 12,000 ×g at 4°C for 20 minutes. The supernatant fractions were pooled and layered onto the top of a cesium chloride step gradient consisting of a 10 mL bottom layer of 1.5 g/mL CsCl and a 10 mL top layer of 1.2 g/mL CsCl. The gradient was spun in a SW28 rotor at 24,000 rpm for 24 h at 20°C. Following the centrifugation of the step gradient, a single virus band appeared at the interface between the two CsCl layers. The virus was recovered from the step gradient using a syringe to puncture through the side of the tube. The crude virus was centrifuged a second time by placing the CsCl solution containing the virus in an SW55 rotor at 28,000 rpm for 24 h at 20°C. Under these conditions, the CsCl forms its own gradient and the virus was separated into two distinct bands, with the virus banding at ~1.33 g/mL (lower band) and 1.31 g/mL (upper band), respectively. The virus bands were harvested independently and dialyzed against 10 mM Tris, pH 8.0, and 250 mM NaCl to remove the CsCl. The virus was stored either at 4°C or at −80°C, depending on the downstream application.

Samples stored at 4°C were sent to the Electron Microscopy Core Facility at the University of Nebraska Medical Center. Grids used for electron microscopy were 200 mesh copper grids coated with formvar and silicon monoxide (Ted Pella, Inc.). A 10 μL drop of each sample collected from the gradient was placed on the grid for one minute. The excess liquid was wicked off with a piece of Whatman 50 filter paper, and the grid was allowed to air-dry for two minutes. A droplet of Nano-Van solution (Nanoprobes, Inc.) was placed on the grid for one minute. The excess liquid was wicked off as above, and the grid was allowed to dry for two minutes. The grids were examined in a Tecnai G2 Transmission Electron Microscope operated at 80 Kv.

RT-qPCR Quantitation of Viral RNA

Viral RNA was quantitated from upper band and lower band virus samples from the isopycnic CsCl gradients (Figure 1) as follows. Virus protein from the two samples was quantitated using the Bradford Assay (BioRad) and equal amounts of virus from the upper and lower bands (5 ng, 2.5 ng, 1.25 ng, 625 pg, and 313 pg) were assayed in duplicate using a SYBR® Green Quantitative RT-qPCR Kit (Sigma) according to the manufacturer's instructions in a 7500 Real Time PCR System (Applied Biosystems). Upper band and lower band virus samples were heated to 95°C for 15 minutes to disrupt particles immediately prior to addition to the RT-qPCR reactions. Data analysis of the RT-qPCR reactions was performed as described by Livak and Schmittgen [9].

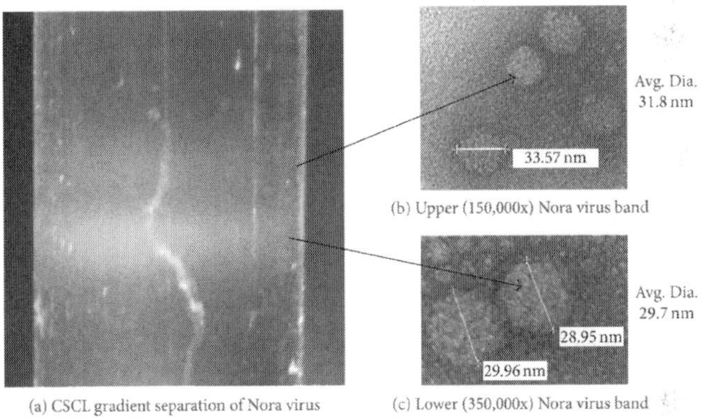

Figure 1. Two distinct sized particles are produced from a Nora virus infection of Drosophila melanogaster witi Rel^{E23} strain flies. Approximately 3,000 Nora virus infected flies were homogenized and processed as described in Section 2. In (a) a representative CsCl gradient is shown, depicting the two bands of virus particles obtained. The bands were collected individually and dialyzed against 10 mM Tris-HCl, pH 8.0, and 250 mM NaCl buffer. Electron micrographs of the particles obtained from the upper band (b) and the lower band (c) are shown. The average sizes of the virus particles were 31.8 nm (upper band; $n = 47$) and 29.7 nm (lower band; $n = 39$), respectively

Cloning and Expression of Nora Virus Genes

Nora virus gene sequences were obtained commercially (GenScript) and were codon optimized for their expression in *E. coli*. An In-Fusion® HD Cloning Kit (Clontech) was used for cloning optimized genes *ORF1*, *ORF3*, *ORF4a*, *ORF4b*, and *ORF4c*, into the linearized *pET28a* vector and they were transformed into competent Stellar cells (Clontech) per the manufacturer's instructions. Colonies were screened by PCR and colonies testing positive for the various Nora virus genes were grown for plasmid DNA preparation using a Qiagen Plasmid preparation kit. Plasmid DNA was used to transform One Shot® BL21 (DE3) pLysS Chemically Competent *E. coli* cells (Invitrogen). These cells were subsequently used for expression of the Nora virus genes.

Transformed E. coli BL21 (DE3) cells were placed in 5 mL of LB-Kanamycin (50 μg/mL) and incubated overnight at 37° C with shaking. Flasks containing 50 mL of LB-Kanamycin broth were inoculated with 500 μL of the overnight cultures and incubated at 37∘ C until midexponential phase ($OD_{600\ nm}$ = 0.5) was reached. Isopropyl β-D-thiogalactoside (IPTG) was added to a final concentration of 1.0 mM to each culture. One-milliliter samples were removed from each culture at 0 hours, 1 hour, 4 hours, and overnight following the IPTG addition. Pellets were collected by centrifugation at 13,000 rpm for three minutes at room temperature and resuspended in 200 μL distilled water, and an equal volume of 2x Laemmli sample buffer (BioRad) was added and boiled for approximately two minutes. Twenty microliter samples were run on Mini-PROTEAN® TGX™ Gels (BioRad) and stained with Coomassie brilliant blue (BioRad). For large scale protein preparations, two milliliters of overnight cultures was placed in 200 mL fresh LB-Kanamycin. Cells were incubated at 37∘ C until midexponential phase ($OD_{600\ nm}$ = 0.5) and IPTG was added to a final concentration of 1.0 mM to each culture and allowed to grow for an additional 24 h at 37∘ C incubation.

Immobilized Metal Ion Affinity Chromatography (IMAC)

E. coli cells were collected from 200 mL cultures (see above) by centrifugation at 8,000 ×g for 10 minutes at 4°C. Supernatant fractions were removed and 10 volumes (g/mL) of lysis/wash buffer (300 mM KCl, 50 mM KH_2PO_4, pH 8.0, 20 mM Imidazole, and 6 M Urea) were added to resuspend the pellets. Lysates were sonicated on ice for 15-second pulses with 35 seconds between pulses until a total amount of sonication time of 4 minutes was obtained. Lysates were centrifuged at 12,000 ×g for 10 minutes at 4°C. Supernatant fractions were removed and filtered through a sterile 0.22 μm syringe filter before placing them on the IMAC column.

Purification of his-tagged proteins (rVP3, rVP4a, rVP4b, rVP4c, or rORF1) was performed for each protein individually through separate

1.0 mL IMAC cartridges under denaturing conditions or for all the proteins together through a 5.0 mL IMAC cartridge under denaturing conditions. A column packed with Profinity™ IMAC (BioRad) resin was equilibrated with 5 column volumes of equilibration/wash buffer 1 (300 mM KCl, 50 mM KH_2PO_4, pH 8.0, 20 mM Imidazole, and 6 M Urea). The cartridge was washed with 6 column volumes of both wash buffer 1 and wash buffer 2 (20 mM Tris, pH 8.0, 50 mM KCl, 20 mM Imidazole, and 2 M Urea). Proteins were eluted with 10 column volumes of elution buffer (20 mM Tris, pH 8.0, 50 mM KCl, 10 mM–250 mM Imidazole, and 2 M Urea) and analyzed by SDS-PAGE with Coomassie blue staining. Purified protein samples were pooled for determination of protein concentration using Pierce™ BCA Protein Assay Kit (Thermo Scientific) following kit instructions.

Preparation of Mouse Antisera against Nora Virus and Recombinant Nora Virus Proteins

All experimental procedures were reviewed and approved by the UNK Institutional Animal Care and Use Committee (IACUC # 091311). Mouse antisera against purified virus or the individual recombinant proteins were prepared as follows. CD-1 outbred retired breeder female mice (Charles Rivers) were used for production of antisera. Mice were housed, 5 per cage, in a holding facility maintained at constant temperature (25°C) under a 12 : 12 light/dark cycle. Food and water were provided *ad libitum*. All mice were prebled via the retroorbital route [10]. Twenty micrograms of virus (lower band) or recombinant protein was mixed with enough 10 mM Tris, pH 8.0, and 250 mM NaCl buffer to give a volume of 100 μL. An equal volume of Freund's complete adjuvant (Pierce Chemical Co.) was added and an emulsion was formed by passing the mixture back and forth through a double-hub needle. Two hundred microliters of the emulsion containing the virus or protein was administered to each mouse via a subcutaneous route on day one. Four weeks later, a test bleed from the mice was taken and a booster injection identical to the primary injection was given, with the exception that the inoculum was prepared using Freund's incomplete adjuvant (Pierce Chemical Co.). Every two weeks following the booster injection, additional test bleeds were taken until the titer waned. In some instances, a tertiary injection was given and additional test bleeds were taken as described.

Western Blot Analysis of Viral Proteins

Viral proteins from whole virus (upper and lower bands), recombinant proteins, or infected cell lysates were run on Mini-PROTEAN® TGX™ Gels (BioRad) as described above and subsequently transferred to nitrocellulose membranes using a semidry blot format (Trans-Blot Turbo Transfer System; BioRad). Following protein transfer, the membranes were blocked with blocking buffer (5% nonfat powdered milk buffered in Tris-HCl, pH 8.0, 150 mM NaCl, and 0.05% Tween 20 (TBST)). The various antisera, either against whole virus or the individual proteins, were diluted in blocking buffer at dilutions varying from 1:500 to 1:2000 and incubated with the membranes for at least 1 h at room temperature or overnight at 4°C. Following incubation with the primary antibody, the membranes were washed three times for 10 minutes each with TBST. After washing, the secondary antibody (goat anti-mouse IgG alkaline phosphatase conjugate; Pierce Chemical Co.) was diluted in blocking buffer at 1:10,000 and incubated with the membrane for at least 1 h at room temperature. The membranes were washed twice with TBS and developed using NBT/BCIP (Pierce Chemical Co.). Blot development was stopped after several minutes with 2 rinses of distilled water.

RESULTS

Purified Nora Virus from Infected Flies Produces Two Types of Particles

As a first step to fully characterizing the protein composition of Nora virus, we modified the Nora virus purification procedure [2] to include an isopycnic CsCl gradient that followed the CsCl step gradient (see Section 2). The results obtained from the isopycnic gradient (Figure 1) yielded two virus bands. The lower band (ρ = 1.33 g/mL) was previously reported [2]. The upper band (ρ = 1.31 g/mL) typically contained less virus particles than the lower band as determined by protein assay. When the particles from each band were observed independently by electron microscopy, the average particle sizes were different. The lower band virus averaged a diameter of

29.7 nm ($n = 47$) and the upper band averaged a diameter of 31.8 nm ($n = 39$).

One potential reason the two particle sizes were observed was the upper band virus particles represented empty capsids. This could result in a larger particle size because the RNA may assist the particles to condense upon assembly as reported for other RNA viruses [11]. It would also account for the difference in density of the upper band particles. Using RT-qPCR on the two particles would provide a means of quantitating the amount of RNA in each band. To this end, virus protein was quantitated via Bradford Assay and equal amounts of protein from the two virus bands were used in RT-qPCR reactions to subsequently quantify the amount of RNA present in the two particle types. Viral RNA (data not shown) is approximately 12-fold more abundant in the lower band particles. This result is consistent with the upper band being primarily composed of empty capsids and the lower band representing complete virions.

Expression and Purification of Recombinant Nora Proteins

Monospecific antisera to the individual viral proteins would allow the characterization of the protein composition of the two particle types. To accomplish this, each individual viral gene, except *ORF2*, was amplified by PCR from codon optimized genes (see Section 2). After each PCR product was obtained in purified form, they were added to *pET28a* linearized vector and ligated together using an In-Fusion® HD Cloning Kit (Clontech). The subsequent plasmids were purified and transformed into BL21 DE3 *E. coli* cells for expression. The proteins were purified from the *E. coli* cultures using the histidine tag added to them by sequences from the *pET28a* vector. The results of the purification are shown in Figure 2. With the exception of VP3, all the viral proteins purified as a single purified species as determined by Coomassie blue-stained SDS-PAGE gels. The lighter band in the VP3 lane appears to be a host *E. coli* protein, as it was not detected by antiserum prepared against whole virus particles (Figure 3). All the virus proteins migrated in accordance with their predicted sizes.

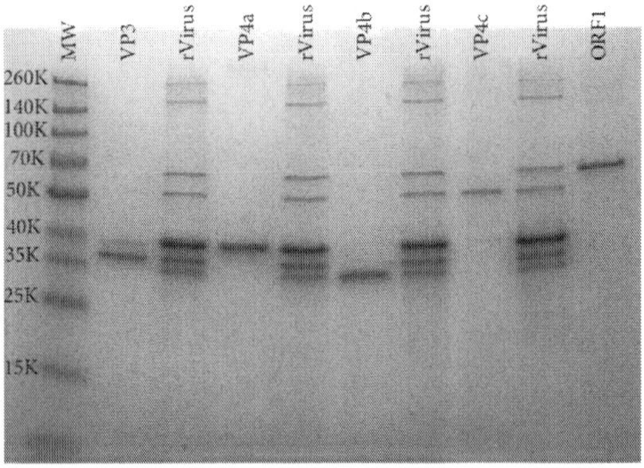

Figure 2. SDS-PAGE analysis of recombinant Nora virus proteins (rNVPs). Recombinant Nora virus proteins were produced by overexpression in *E. coli* cells as described in Section 2. The resultant proteins were purified using Ni^{+2} charged IMAC columns either as individual proteins (lanes labeled VP3, VP4a, VP4b, VP4c, and ORF1) or as a mixture of the 5 proteins (lanes labeled rVirus). The proteins were subsequently separated on 10% SDS-PAGE gels followed by Coomassie blue staining. The results showed the relative purity of the preparations and demonstrated that the proteins readily copurified when mixed together.

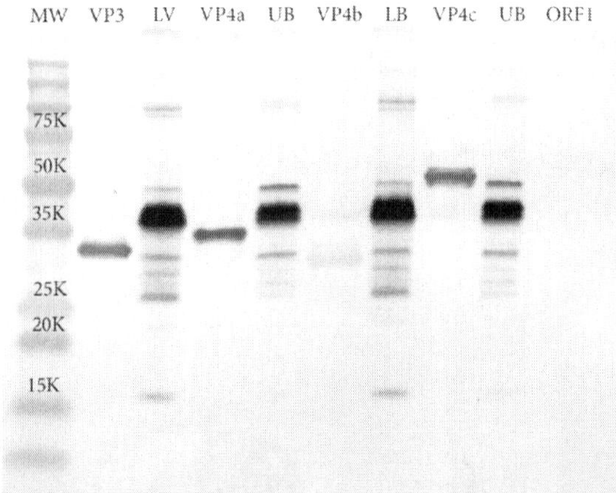

Figure 3. Western blot analysis of Nora virus proteins and recombinant proteins. Mouse antiserum was prepared against whole Nora virus particles (lower band virus particles, see Figure 1) as described in Section 2. Nora virus purified on CsCl gradients was separated into upper band particles (virus UB) and lower

band particles (virus LB). The viral protein components of the UB and LB particles were separated on 10% SDS-PAGE gels and blotted onto nitrocellulose membranes, along with the *E. coli* expressed recombinant Nora virus proteins (VP3, VP4a, VP4b, VP4c, and ORF1). The recombinant proteins were either strongly detected (VP3, VP4a, and VP4c), weakly detected (VP4b), or not detected at all (ORF1). There does not appear to be any discernable polypeptide composition differences between the upper and lower band virus particles.

If all five virus proteins were mixed together prior to column purification, they all copurified. This was expected. However, this did result in some higher molecular weight species (~160 K and 260 K) that were not observed from the individual viral protein purifications. We suggest these may be complexes of individual or multiple viral proteins, as they are detected by whole virus antiserum in Western blots (see Figure 3). In addition, these complexes resisted the denaturation steps of SDS-PAGE, indicating a very tight association.

Western Blot Analysis of the Two Particle Types Shows They Are Identical in Protein Composition

The protein composition of the upper and lower band virus particles was further characterized by preparing antisera against a mixture of these particles and subsequently performing a Western blot analysis (Figure 3). In the lanes labeled virus LB and virus UB (lower band and upper band virus particles, resp.) no discernible differences could be observed in polypeptide composition. All the protein species previously reported [1] were detected in both particle types. Therefore, the difference in particle size is again suggested to be from the lack of genomic RNA in the upper band particles.

The antisera made against whole virus particles were shown to react strongly with each of the recombinant-produced proteins, with the exception of VP4b (weak reaction) and ORF1 (no detectable reaction). It is not clear why VP4b gave a weak reaction, as the monospecific sera made against VP4b react very strongly with purified virus (Figure 4). This may indicate that VP4b is not immunogenic or it is not sufficiently exposed in the virion. ORF1 protein was predicted not to react with whole virus antisera as it was not previously reported to be part of the virus particle [1].

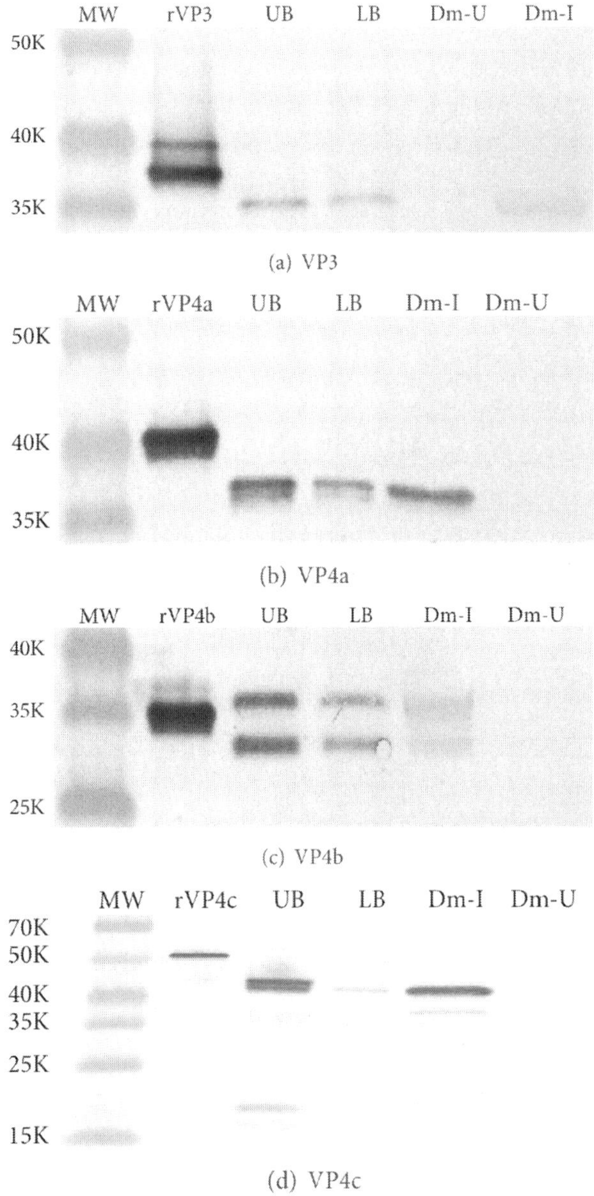

Figure 4. Western blot analysis of the polypeptide composition of the upper band and lower band Nora virus particles using monospecific antisera. Mice were hyperimmunized with rVP3, rVP4a, rVP4b, or rVP4c and the resultant sera were collected as described in Section 2. (a) shows the results of using anti-VP3 antiserum. An SDS-PAGE gel was loaded with molecular weight markers

(MW), recombinant VP3 (rVP3), 5.55 μg of upper band virus particles (UB), 5.55 μg of lower band virus particles (LB), Nora virus infected fly lysate (Dm-I), or Nora virus uninfected fly lysate (Dm-U). After electrophoretic separation, the polypeptides were blotted onto a nitrocellulose membrane and developed as described in Figure 4 using a 1 : 1,000 dilution of the primary antibody. (b) (VP4a), (c) (VP4b), and (d) (VP4c) were all run identically to (a) with the exception of the dilution of the primary antibody which was 1 : 500, 1 : 2,000, and 1 : 500, for VP4a, VP4b, and VP4c, respectively.

Protein Composition of Nora Virus Particles Consists of at Least Six Polypeptides

As a final step to characterize the protein composition of the two types of Nora virus particles, antisera were prepared against each of the recombinant proteins (rVP3, rVP4a, rVP4b, rVP4c, and rORF1). Western blots were developed with each of the antisera (Figure 4). The samples run on each blot were the individual recombinant proteins, upper (UB) and lower (LB) band Nora virus particles, and *D. melanogaster* infected (Dm-I) or uninfected (Dm-U) cell lysates. In each case, the recombinant proteins migrate at a larger size than those from purified virus or infected cells do. This is due to the presence of a HIS-tag on the recombinant proteins. The exception to this is VP4b. The protein from infected cells and purified virus migrated at a larger size than the rVP4b. We suggest VP4b may be posttranslationally modified in some as yet uncharacterized manner, increasing the predicted molecular weight of the protein.

ORF1 monospecific antiserum did not detect proteins in purified virus particles but did detect a single protein species in infected cells (data not shown). This result was consistent with previous reports of Nora virus virion protein composition [1]. As predicted, the VP3 antiserum (Figure 4(a)) detected a single protein species in the virus particles at ~35 K in molecular weight. An identical protein species was detected in infected cell lysate, but not in uninfected cell lysate. The monospecific serum prepared against VP4a (Figure 4(b)) detected two protein species in both virus particle types at ~37 and ~36 K, whereas, in infected cell lysates, only the 37 K protein species was detected. This may suggest that the 37 K protein was processed upon virus assembly or is an artifact of virus purification. Figure 4(c) shows the results obtained using monospecific antiserum against VP4b. Two protein species were detected, migrating at ~36 K and ~30 K in molecular weight, in both virus particle types and in infected cell lysates. The migration pattern of both species of VP4b suggests both are modified. The 30 K species was previously reported to be derived by a C-terminal proteolytic cleavage from the 36 K species [1]. However, the 36 K species also appears to be modified in some way, as it migrates at a larger size than the recombinant protein (~35 K). The nature of this modification has not been investigated. VP4c antiserum yielded the most complex pattern of the five proteins tested (Figure 4(d)). The upper band virus particles showed at least 5 protein species detected (~42 K, ~41 K, ~35 K, ~20 K, and ~18 K). The blot shown detected only the ~41 K polypeptide in the lower band particles, but in other blots (data not shown) where the NBT/BCIP substrate incubation time was extended to as much as 30 minutes, the other species did stain, but background levels of protein detection in the cell lysate lanes increased dramatically. Two of the protein species were readily detected in the infected cell lysates (~41 K and ~35 K), but the other protein species were not routinely seen. Again, this may be due to processing of viral proteins upon release of mature virus particles, or the additional proteins may be due to an artifact of virus purification. The sizes and numbers of protein species were similar to those previously described [1].

DISCUSSION

An initial step necessary to understand the function of the protein components of a virus is to determine precisely what proteins are virion components and what proteins specified by the genome are nonstructural proteins. The results of this study have clarified this for the *D. melanogaster* Nora virus. Most insect RNA virus virions are composed of relatively few proteins. For example, CrPV is composed of 4 capsid proteins of molecular weights 33, 31, 30, and 8 kDa [12]. *Drosophila* C virions consist of 5 viral proteins, again in the size range of CrPV [13]. Many other RNA viruses of insects also exhibit similar polypeptide composition (reviewed in [14]). The previously reported polypeptide composition of Nora virus [2] suggests a more complex capsid structure consisting of at least 9 polypeptide species ranging in molecular weight from 48 kDa to 17 kDa. The viral genome is predicted to specify 4 structural proteins: one encoded in *ORF3* and the other 3 are derived from *ORF4* via proteolytic cleavage of a polyprotein precursor [1]. This indicated one of two possibilities: either the viral proteins undergo extensive posttranslational processing, primarily proteolytic in nature, or the number of protein species is an artifact of virus purification. If intracellular proteolytic processing was involved, then the protein components found in the mature virions should also be found in the infected cell. On the other hand, if the polypeptide pattern observed in purified virus is an artifact of virus purification, then no such correlation should be found.

The purification procedure for Nora virus was modified from that previously reported [2] to include an extra CsCl gradient step (Figure 1). This yielded upper and lower virus bands of different densities (1.31 and 1.33 g/mL) and sizes (31.8 nm and 29.7 nm). The viral polypeptide composition was identical between the two particles (Figure 3), indicating that the difference in both size and density may lie with the presence or absence of genomic RNA. This was confirmed by RT-qPCR, showing that the lower band contains approximately 12-fold more RNA than does the particle population in the upper band. We suggest that the difference in particle size is due to the viral proteins condensing around the viral RNA in the lower band particles as is known to occur for other RNA viruses during assembly [11]. The detection of two particles that are present as complete virions or empty particles was not unexpected as insect viruses often make much more structural protein than is needed to encapsidate the amount of viral RNA that is replicated (reviewed in [14]).

The production of antisera against whole virus and against the individual viral proteins was used to determine the protein composition of the Nora virus particles. Antibodies against purified virus did not detect ORF1 in virus particles (Figure 3, far right hand lane), nor was it detected in virus particles by monospecific sera against ORF1 protein (data not shown) and therefore this protein is almost certainly not part of the virion. This is consistent with other reports of Nora virus protein composition [1].

Antiserum prepared against VP4a (Figure 4(b)) detected two protein species in virus particles at ~37 and ~36 K, but only the 37 K species was detected in infected cells. This suggests that the 36 K species either is an artifact of virus purification or is derived proteolytically from the 37 K protein during virion release. VP4b was shown previously to exist in two forms [1], a 32 K species and a 26 K species. The 26 K species is a proteolytic cleavage product of the 32 K species. Our results are consistent with this result (Figure 4(c)). Analysis of the amino acid sequence shows a phenylalanine residue 21 amino acids from the C-terminus. This is the cleavage site indicated by Ekström et al. [1] to produce the 26 K species and therefore a chymotrypsin-like protease would be predicted to perform this processing step. The size differential between our results and those reported by Ekström et al. [1] is most likely due to their incorporation of Urea into their SDS-PAGE gels. Unexpectedly, the recombinant protein representing the VP4b sequence migrated at a smaller size than did the virion protein equivalent. This should not be the case as the recombinant protein contains extra amino acids in the form of a histidine purification tag. We therefore suggest that the VP4b species might be modified in some additional fashion during or after translation. Ekström et al. [1] reported the number of proteins originating from the VP4c protein to be at least 5 processed polypeptides. We also detected 5 protein species (Figure 4(d)) of sizes consistent with this observation in purified virus. However, in infected cells, only two protein species were readily detected, the 41 K and 35 K species. Again, this may suggest that the other species derived from VP4c may be an artifact of purification or are produced during virus assembly and release. While analysis of the amino acid sequence of VP4c shows several potential chymotrypsin and trypsin cleavage sites that could yield the 35 K species, without mass spectrometry data, we cannot be certain where the precise cleavage site resides. VP3 monospecific antisera consistently detected only a single viral protein, regardless of whether it was from virus or from infected cells (Figure 4(a)).

CONCLUSION

As indicated previously, insect RNA viruses are typically simple in terms of protein composition. We therefore deem it unlikely that Nora virus is composed of such a complex polypeptide mixture. With a particle size of only 31 nm, this would imply a low copy number of each of these proteins, and this would make assembly much more complex than is typical of picornaviruses or picorna-like viruses (reviewed in [7]). We therefore suggest, based on the Western blot results from the infected cell, that the number of proteins composing the Nora virus virion is less than the 9 previously indicated [1] and that some of those protein species are produced by protease action as an artifact of virus purification. The fact that some of the viral proteins, especially VP4c, are susceptible to protease action is demonstrated by the difference in polypeptide composition of virus purified from infected flies versus virus purified from feces [1]. A cell culture system that could be used to propagate Nora virus is desperately needed to be able to investigate these issues further. At the present time, we suggest that Nora virus particles are composed of 6 proteins: VP3, VP4a, two species of VP4b (36 K and 30 K), and two species of VP4c (41 K and 35 K). This polypeptide composition would be more in line with other picornavirus-like RNA insect viruses such as CrPV [12] and *Drosophila* C virus [13]. However, we cannot rule out the possibility that proteolytic cleavage occurs as a result of virus release, making the polypeptide composition of Nora virus virions more complex than these other viruses. Further studies needed to define these maturation events await a cell culture system that is able to propagate Nora virus.

COMPETING INTERESTS

The authors declare that they have no competing financial interests.

AUTHORS' CONTRIBUTIONS

All authors contributed to this paper.

ACKNOWLEDGMENTS

The authors thank Drs. Hultmark and Ekström for their gift of the original Nora virus stocks. In addition, they thank Angela Wang, Andrew Prososki, Morgan Abert, Kirsten Lipps, Tad Fuchs, and Alexis Page for technical help and Tom Bargar of the UNMC Electron Microscopy Center for his work. The research was supported by grants to K. A. Carlson from the National Center for Research Resources (5P20RR016469) and the National Institute for General Medical Science (8P20GM103427), a component of the National Institutes of Health.

REFERENCES

1. J.-O. Ekström, M. S. Habayeb, V. Srivastava, T. Kieselbach, G. Wingsle, and D. Hultmark, "*Drosophila* Nora virus capsid proteins differ from those of other picorna-like viruses," *Virus Research*, vol. 160, no. 1-2, pp. 51–58, 2011.
2. M. S. Habayeb, S. K. Ekengren, and D. Hultmark, "Nora virus, a persistent virus in *Drosophila*, defines a new picorna-like virus family," *Journal of General Virology*, vol. 87, no. 10, pp. 3045–3051, 2006.
3. S. H. Merkling and R. P. Van Rij, "Analysis of resistance and tolerance to virus infection in Drosophila," *Nature Protocols*, vol. 10, no. 7, pp. 1084–1097, 2015.
4. K. Falk, V. Aspehaug, R. Vlasak, and C. Endresen, "Identification and characterization of viral structural proteins of infectious salmon anemia virus," *Journal of Virology*, vol. 78, no. 6, pp. 3063–3071, 2004.
5. L. Liu, C. C. P. Celma, and P. Roy, "Rift Valley fever virus structural proteins: expression, characterization and assembly of recombinant proteins," *Virology Journal*, vol. 5, pp. 82–94, 2008.
6. S. U. Nielsen, M. F. Bassendine, A. D. Burt, D. J. Bevitt, and G. L. Toms, "Characterization of the genome and structural proteins of hepatitis C virus resolved from infected human liver," *Journal of General Virology*, vol. 85, no. 6, pp. 1497–1507, 2004.
7. B. Simizu, K. Yamamoto, K. Hashimoto, and T. Ogata, "Structural proteins of Chikungunya virus," *Journal of Virology*, vol. 51, no. 1, pp. 254–258, 1984.
8. E. J. Cordes, K. D. Licking-Murray, and K. A. Carlson, "Differential gene expression related to Nora virus infection of *Drosophila melanogaster*," *Virus Research*, vol. 175, no. 2, pp. 95–100, 2013.
9. K. J. Livak and T. D. Schmittgen, "Analysis of relative gene expression data using real-time quantitative PCR and the 2(−Delta Delta C(T)) method," *Methods*, vol. 25, no. 4, pp. 402–408, 2001.
10. J. S. Garvey, N. E. Cremer, and D. H. Sussdorf, *Methods in Immunology*, W.A. Benjamin, 3rd edition, 1977.
11. A. Borodavka, R. Tuma, and P. G. Stockley, "A two-stage mechanism of viral RNA compaction revealed by single molecule fluorescence," *RNA Biology*, vol. 10, no. 4, pp. 481–489, 2013.

12. N. F. Moore, A. Kearns, and J. S. K. Pullin, "Characterization of cricket paralysis virus-induced polypeptides in *Drosophila* cells," *Journal of Virology*, vol. 33, no. 1, pp. 1–9, 1980.
13. F. X. Jousset, M. Bergoin, and B. Revet, "Characterization of the *Drosophila* C virus," *Journal of General Virology*, vol. 34, no. 2, pp. 269–283, 1977.
14. J. Xu and S. Cherry, "Viruses and antiviral immunity in *Drosophila*," *Developmental and Comparative Immunology*, vol. 42, no. 1, pp. 67–84, 2014.

10

Sensitive Detection of Fluorescence in Western Blotting by Merging Images

Yukari Kondo[1], Shinichiro Higa[1,2], Takeshi Iwasaki[1], Tomoya Matsumoto[3], Kazumitsu Maehara[1], Akihito Harada[1], Yoshihiro Baba[4], Masatoshi Fujita[2], Yasuyuki Ohkawa[1]

[1] Division of Transcriptomics, Medical Institute of Bioregulation, Kyushu University, Fukuoka, Japan
[2] Department of Cellular Biochemistry, Graduate School of Pharmaceutical Sciences, Kyushu University, Fukuoka, Japan
[3] Sales Department, Microscope Division, Keyence, Osaka, Japan
[4] Division of Immunology and Genome Biology, Medical Institute of Bioregulation, Kyushu University, Fukuoka, Japan

ABSTRACT

The western blotting technique is widely used to analyze protein expression levels and protein molecular weight. The chemiluminescence method is mainly used for detection due to its high sensitivity and ease of manipulation,

Citation: Sensitive detection of fluorescence in western blotting by merging images. Kondo Y, Higa S, Iwasaki T, Matsumoto T, Maehara K, Harada A, et al. (2018) PLoS ONE 13(1): e0191532. https://doi.org/10.1371/journal.pone.0191532

but it is unsuitable for detailed analyses because it cannot be used to detect multiple proteins simultaneously. Recently, more attention has been paid to the fluorescence detection method because it is more quantitative and is suitable for the detection of multiple proteins simultaneously. However, fluorescence detection can be limited by poor image resolution and low detection sensitivity. Here, we describe a method to detect fluorescence in western blots using fluorescence microscopy to obtain high-resolution images. In this method, filters and fluorescent dyes are optimized to enhance detection sensitivity to a level similar to that of the chemiluminescence method.

INTRODUCTION

Cells are regulated according to the central dogma, which is the flow from genetic information to protein expression. The expression levels of proteins dictate cell fate; for example, the expression levels of certain transcription factors regulate skeletal muscle differentiation [1]. Thus, it is critical to quantify protein expression to understand cellular phenomena or potential at the molecular level. Western blotting is the standard method to quantify protein expression [2]. The chemiluminescence method is generally used to detect proteins because of its high sensitivity and ease of use [3, 4]. However, recent developments in proteomics have focused on technologies to detect the expression of multiple proteins simultaneously, as opposed to the chemiluminescence method that detects single proteins. One method for simultaneous detection of multiple proteins which allows more quantitative analysis is fluorescent western blotting [5–7]. However, the detection sensitivity of this method is lower than that of the chemiluminescence method.

In this study, to improve the resolution and sensitivity of fluorescent western blotting, we focused on the fluorescence microscopy step. The microscope's CCD camera is capable of capturing high-resolution images, but it has a small visual field. Thus, we captured an image with a high resolution and a wide field by merging multiple fluorescence micrographs. In addition, we successfully increased the detection sensitivity to a level comparable to that of the chemiluminescence method by optimizing the filters and fluorescent dyes.

MATERIALS AND METHODS

Cells

The mouse C2C12 myoblast cells (CRL-1772, ATCC) were cultured in Dulbecco's modified Eagle's medium (DMEM) containing 20% fetal bovine serum, and were harvested at about 70% confluency. Differentiated cells were transferred to DMEM medium containing 2% (v/v) horse serum and cultured for 24 or 72 hours before analyses.

Immunoblotting

Purified recombinant GST was denatured in bromophenol blue (BPB)-free 2×sodium dodecylsulphate (SDS) sample buffer at 95°C for 5 min. Cells were washed twice in phosphate buffered saline (PBS), resuspended in BPB-free 2×SDS sample buffer, and denatured at 99°C for 2 min. The samples were separated by SDS-PAGE (12% acrylamide), and transferred to an Immun-Blot® LF PVDF Membrane (BioRad, Hercules, CA, USA). The membrane was blocked with 5% skim milk powder in Tris-buffered saline containing 0.05% Tween 20 (TBST), then incubated with primary antibodies in Hikari Solution A (Nacalai Tesque Inc., Kyoto, Japan), and subsequently incubated with secondary antibodies in TBST. The primary antibodies included rabbit monoclonal anti-GST (EPR4236, Abcam, Cambridge, MA, USA; 1:1000), rabbit monoclonal anti-desmin (Y66, Abcam; 1:1000), rat monoclonal anti-Hsp90 alpha (4H7A5, Kishida, Osaka, Japan; 1:2500) and mouse monoclonal anti-myogenin (F5D, Santa Cruz Biotechnology, Santa Cruz, CA, USA; 1:1000). The secondary antibodies included horseradish peroxidase-conjugated anti-rabbit IgG antibody (GE Healthcare Life Science, Buckinghamshire, UK; 1:5000), CF™488A goat anti-rabbit IgG (H+L) (Biotium, Hayward, CA, USA; 1:5000), Alexa Fluor™ 546 goat anti-rabbit IgG (Thermo Fisher Scientific, Rockford, IL, USA; 1:5000), Alexa Fluor™ 647 goat anti-rabbit IgG (Thermo Fisher Scientific, 1:5000), Alexa Fluor™ plus 488 goat anti-rabbit IgG (Thermo Fisher Scientific, 1:5000), Alexa Fluor™ plus 555 goat anti-rabbit IgG (Thermo Fisher Scientific, 1:5000), Alexa Fluor™ 546 goat anti-mouse IgG (H+L) (Thermo Fisher Scientific, 1:5000), ECL Plex goat-α-rabbit IgG, Cy5 (GE Healthcare Life Science, 1:5000), and Alexa Fluor™ 647 chicken anti-rat IgG (H+L) (Thermo Fisher Scientific, 1:5000). Chemiluminescence was analyzed using Chemi-Lumi One Ultra (Nacalai Tesque Inc.).

Visualization and Data Analysis

The membrane was placed on a black acrylic plate and covered with cover glass (85 × 125 mm, No. 1-S, Matsunami Glass, Osaka, Japan). Images were visualized under a fluorescent microscope (BZ-X710, Keyence, Osaka, Japan) with a replaced objective lens (CFI Plan Apochromat λ 4×, Nikon Instruments Inc., Tokyo, Japan) and joined using its software BZ-analyzer (Keyence). The fluorescence microscope was equipped with a motorized stage and a CCD camera with the dynamic range of 14 bit, and excitation light emission was kept at the minimal level (low-photobleach mode). The filters included BZ-X filter GFP (OP-87763, Keyence), BZ-X filter Cy5 (OP-87766, Keyence), BZ-X filter TRITC (OP-87764, Keyence), ET-Narrow Band EGFP to minimize autofluorescence (49020, Chroma Technology, Brattleboro, VT, USA), ET-Cy5 narrow excitation (49009, Chroma), and ET-Gold FISH (49304, Chroma). Images were optimized using Adobe PhotoShop CS6. Quantification was performed using ImageJ and relative integrated density was calculated as follows: (area of GST × mean GST)–(area × mean of average of two control areas).

RESULTS

Joining Images to Capture High-Resolution Fluorescent Western Blot Image

To perform fluorescent western blotting with high resolution and high detection sensitivity, we captured high-resolution images under a fluorescence microscope and merged those images. Microscopic images (8.8 × 6.6 mm) overlapping by 30% of the image on each side were captured in three rows of 29, and then joined and shade corrected automatically using the BZ-analyzer software. As a result, we successfully captured an image with a wide field, high resolution, a broad area, and sensitivity comparable to that of the chemiluminescence method (Fig 1). Because we used the low-photobleach mode with short-time exposure, fluorescence signal was mostly not repressed. However, the captured image contained noise signals resulting from membrane autofluorescence, leakage of excitation light, nonspecific signals and transmitted scattering light.

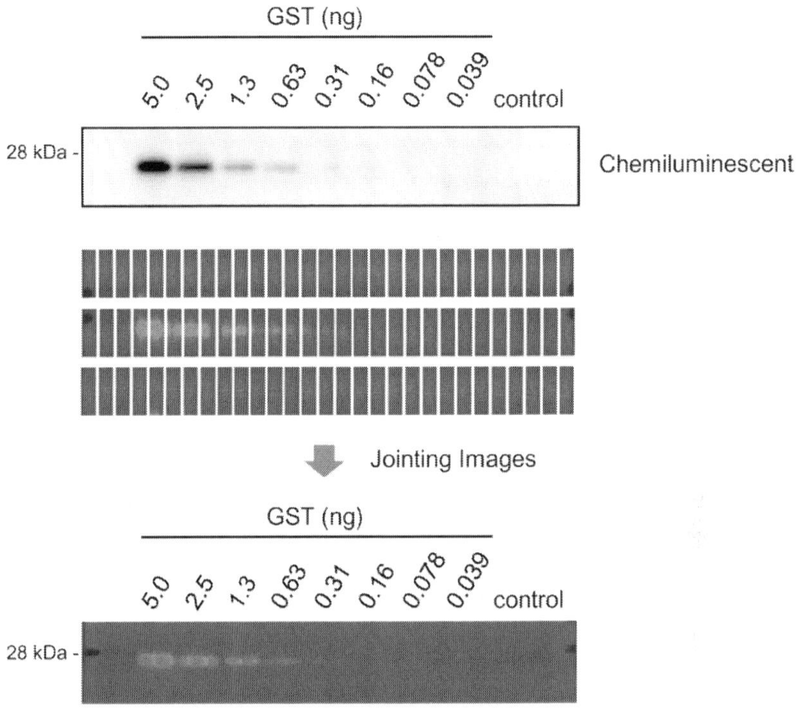

Figure 1. Joined image generated using fluorescence detection method.

Dilution series of recombinant GST detected by our fluorescence method and by the chemiluminescence method. In our method, multiple images are captured using fluorescence microscopy, and joined to make a single image with a wide field and high resolution. The joined image has a broad area and high sensitivity comparable to that of the chemiluminescence method.

Narrowing Transmission Wavelength Range To Enhance Detection Sensitivity

To reduce the noise signals and enhance detection sensitivity, we used two approaches to optimize filters and fluorescent dyes. First, we optimized the filter to remove autofluorescence and separate it from excitation light. The excitation and emission filters used for capturing tissue images have wide transmission wavelength ranges to detect, signals from a wide wavelength range, but undesirable signals are also transmitted and can result in noise. Narrowing the transmission wavelength range of the filter can remove unnecessary signals to enhance detection sensitivity, but there is the risk of weakening the target signal intensity. Thus, we investigated whether noise was removed and the target signal intensity was maintained by using a filter with a narrower transmission wavelength range. We serially diluted recombinant GST, and labelled it with CF488A (green), Alexa 546 (yellow), and Alexa 647 (far-red). Each signal was detected using the general filter and the narrow-band filter, and the relative integrated density of GST was calculated. The relative integrated density of 1.56 ng GST is shown in Fig 2. In the green wavelength region (495–570 nm), the relative integrated density of GST was approximately doubled by using the narrow-band filter. The relative integrated density of GST was similarly enhanced in the yellow (570–590 nm) and far-red (710–850 nm) wavelength regions. These results indicated that removal of noise by narrowing the transmission wavelength range of the filter significantly enhanced the detection sensitivity.

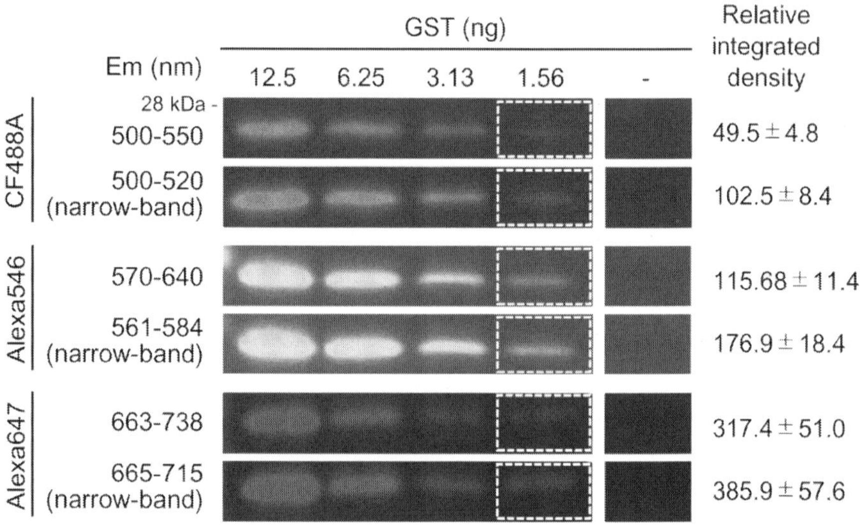

Figure 2. Reduction in noise signals by narrowing emission wavelength region.

Effects of using narrow-band filters on detection sensitivity. Joined images of dilution series of GST and background area are shown. GST was visualized using anti-GST antibody and fluorescent dyes conjugated to secondary antibody (CF488A, Alexa 546, and Alexa 647). Emission wavelength (Em) of filters and relative integrated density of 1.56 ng GST using general filters (upper lane) and narrow-band filters (lower lane) are shown (mean ± standard error of three independent experiments). Exposure time was 1/3 seconds in the normal filter and 1/7.5 seconds in the narrow-band filter in detecting GFP and Alexa 546, and was 1–1.5 seconds in the normal filter and 1.5 seconds in the narrow-band filter in detecting Alexa 647.

Use of High Fluorescence Intensity Dyes To Enhance Detection Sensitivity

Since the use of a narrow-band filter may decrease the target signal, we selected dyes with high fluorescence intensity to enhance the target signal. The fluorescence intensity is increased relative to the product of the molar extinction coefficient (ε) and fluorescence quantum yield (φ). To select fluorescent dyes that would enhance the signal, we compared the signal intensity of GST using dyes with different ε and φ in each wavelength range (Fig 3). In the green wavelength region, the relative integrated density of 1.56 ng GST was higher with CFF488A than with Alexa plus 488. In the yellow wavelength region, the relative integrated density of GST was higher with Alexa 546 than with Alexa plus 555, and in the far-red wavelength region, the relative integrated density of GST was higher with Alexa 647 than with Cy5. The product of the ε and φ of Cy5 is 75,000, and that of Alexa 647 is 89,100. This value may be used as an index to select a fluorescent dye, because those with higher values tended to show higher relative integrated density of GST in each wavelength region. These results confirmed that selection of an appropriate fluorescent dye enhanced the target signal and increased the detection sensitivity.

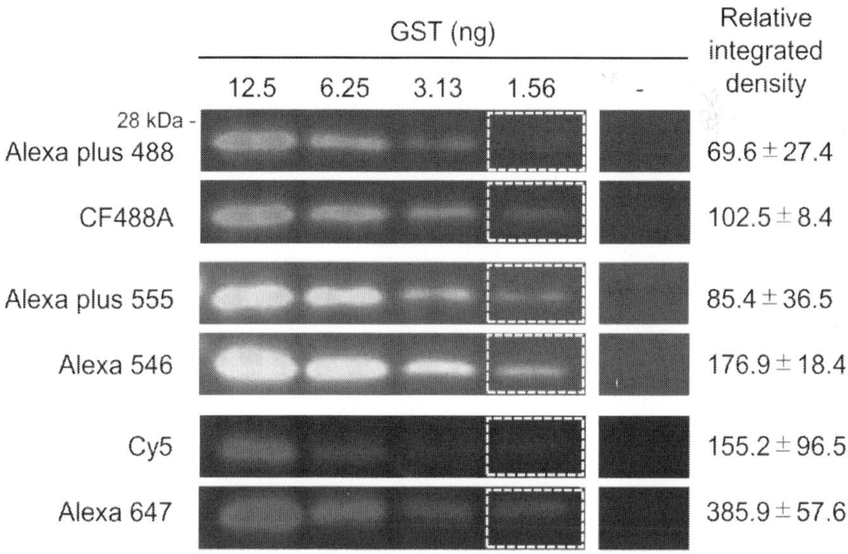

Figure 3. Increase in signal intensity by selecting appropriate fluorescent dyes.

Comparison of detection sensitivity among fluorescent dyes. Joined images of dilution series of GST and background area are shown. GST was detected by anti-GST antibody, fluorescent dye conjugated to secondary antibody, and narrow-band filters (Em: 500–520 nm, 570–590 nm, and 665–715 nm). Relative integrated density of 1.56 ng GST signal is shown (mean ± standard error of three independent experiments). Exposure time was 1/1.5 seconds, 1/3 seconds and 1 second in green, yellow and far-red wavelength regions respectively.

Quantitative and Simultaneous Detection Of Multi-Fluorescence Signals

To confirm the practicality of merging fluorescence micrographs, we demonstrated the simultaneous detection of three proteins using optimized filters and fluorescent dyes (Table 1). With these optimized filter-dye combinations, the relative integrated density was linearly increased relative to amount of loaded GST (S1 Fig). Specifically, we detected desmin (muscle-specific intermediate filament), myogenin (muscle-specific transcription factor) and Hsp90 (molecular chaperone), which showed different expression patterns during skeletal muscle differentiation in C2C12 myoblast cells. Desmin expression increases upon induction of differentiation [8], myogenin is not expressed under normal growth conditions but increases after differentiation [9], and Hsp90 is expressed continuously [9]. Consistent with known expression patterns, the relative expression level of desmin was increased to 1.2 in 24 hours and to 1.9 in 72 hours after upon induction of differentiation. Myogenin was not expressed before differentiation, but its relative expression level increased to 4.6 in 24 hours and to 3.2 in 72 hours after differentiation, and Hsp90 was stably expressed. There was some antibody-derived noise in the detection of myogenin and Hsp90, but our results were consistent with those of previous reports (Fig 4). These results confirmed that multiple proteins could be detected concurrently and quantitatively by this fluorescence detection method using a fluorescence microscope.

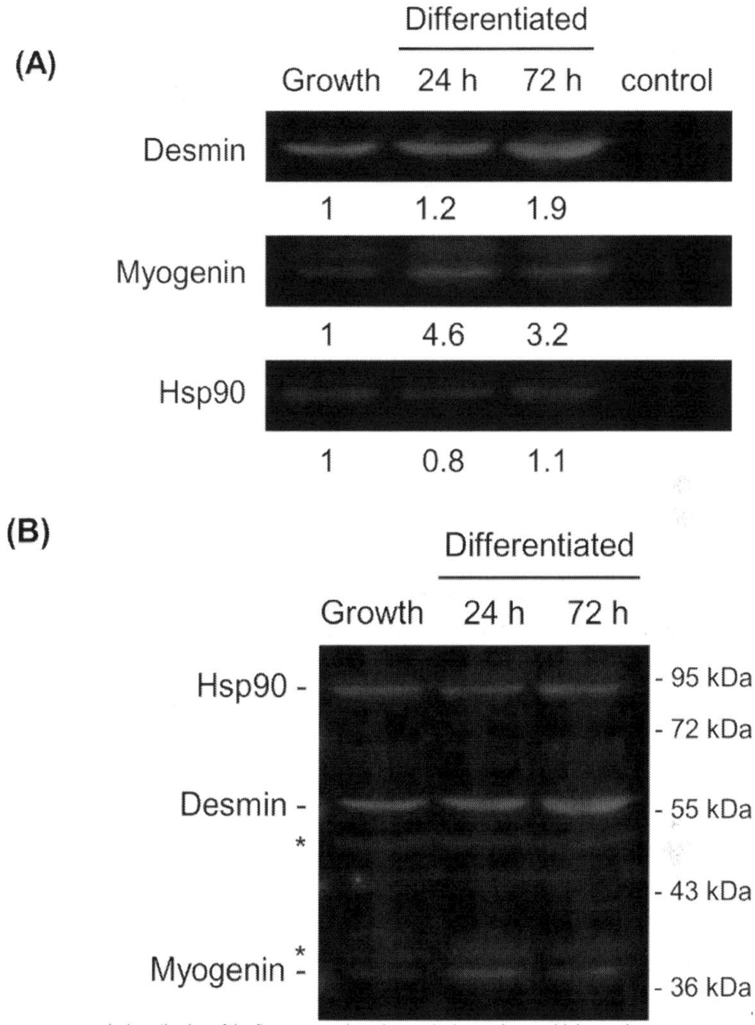

Figure 4. Practical application of the fluorescence detection method to analyze multiple proteins.

(A) Joined images of fluorescent western blot of C2C12 myoblast cells under normal growth and differentiated conditions. Control lane contained sample buffer as negative control. Desmin (green), myogenin (red), and Hsp90 (blue) are displayed as false colors. Relative integrated density values normalized to the values at the growth condition are shown. Exposure time was 1/40 seconds in desmin, 1/8.5 seconds in myogenin and 1/5 seconds in Hsp90. (B) Merged images described in (A). Asterisk shows nonspecific band.

Table 1. Optimized combinations of filter and fluorescence dye in each color wavelength region.

	Filter (Em)	Fluorescence dye
Green (495–570 nm)	Narrow-band (500–520 nm)	CF488A
Yellow (570–590 nm)	Narrow-band (561–584 nm)	Alexa546
Far-red (710–850 nm)	Narrow-band (665–715 nm)	Alexa647

Optimized combinations of filter and fluorescence dye in each color wavelength region (see Fig 4).

https://doi.org/10.1371/journal.pone.0191532.t001

DISCUSSION

The main problems of fluorescence detection methods are low resolution and poor detection sensitivity. In this study, we successfully captured a high-resolution image by merging multiple images. This technique of capturing a wide-field image by merging fluorescence microscope images has been used to visualize immunostaining in various tissues. When merging images captured under the high-power objective lens, image distortion may occur because of objective lens aberration. However, images are captured under low-power objective lens in our method, and so there is little image distortion. We reduced noise and enhanced the signal intensity by using a narrow-band filter and selecting high fluorescence intensity dyes. Consequently, the detection sensitivity was enhanced to a level comparable to that of the chemiluminescence detection method. Therefore, our methods would be suitable for quantitative protein expression analysis with higher resolution and detection sensitivity. In future studies, both the filter and fluorescence dye should be further optimized to increase the color contrast in particular wavelength regions. As our method differs from the chemiluminescence method only in using fluorescent dyes conjugated to the secondary antibody and detecting signal using the fluorescence microscope, it enables fluorescence western blotting at a low installation cost without modifying the chemiluminescence method protocol.

Our method is not limited in the selection of target proteins, antibody host species, fluorescence dyes or filters, and does not require a special detector but only a standard fluorescence microscopy. One advantage of this fluorescent western blotting technique is that multiple proteins can be detected concurrently. Thus, we examined whether multiple proteins were detected using the optimized filter and fluorescent dyes. We detected three kinds of proteins simultaneously and quantitatively in the green, yellow, and far-red wavelength regions. Because the fluorescence microscope has a wide range of excitation wavelengths, we think that it is possible to detect up to five kinds of proteins concurrently by establishing appropriate conditions for detection in the blue and near-infrared wavelength regions. In addition, exposure time per vision is shorter than 1.5 seconds lessening the risk of the photo-bleaching of signals. However, because antibody-derived nonspecific bands may be detected depending on the type of target protein, the results may be complex and difficult to interpret. For exhaustive protein expression analyses, it is necessary to increase the number of useful wavelength regions. It would be useful to develop a technique that is not limited by the type of fluorescent dyes to analyze the expression of many proteins simultaneously.

SUPPORTING INFORMATION

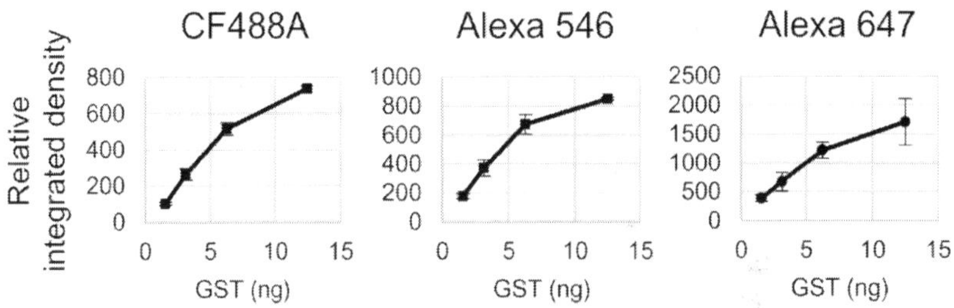

S1 Figure. Correlations between the protein amount and relative integrated density.

The relative integrated density values of GST using narrow-band filters (Em: 500–520 nm, 570–590 nm, and 665–715 nm) and fluorescent dyes (CF488A, Alexa 546, and Alexa 647, respectively) conjugated to the secondary antibody are shown (mean ± standard error of three independent experiments) described in Fig 3. The semi-linear increase slows down after a certain point, which might suggest that fluorescence signal starts to be saturated.

ACKNOWLEDGMENTS

We thank J Nogami and T Komatsu for advice; M Kato, K Fukuyama and Y Hashiguchi for technical support.

REFERENCES

1. Braun T, Gautel M. Transcriptional mechanisms regulating skeletal muscle differentiation, growth and homeostasis. Nat Rev Mol Cell Biol. 2011;12(6):349–61. pmid:21602905
2. Towbin H, Staehelin T, Gordon J. Electrophoretic transfer of proteins from polyacrylamide gels to nitrocellulose sheets: procedure and some applications. Proc Natl Acad Sci U S A. 1979;76(9):4350–4. pmid:388439
3. Bronstein I, Voyta JC, Thorpe GH, Kricka LJ, Armstrong G. Chemiluminescent assay of alkaline phosphatase applied in an ultrasensitive enzyme immunoassay of thyrotropin. Clin Chem. 1989;35(7):1441–6. pmid:2667800
4. Kricka LJ. Chemiluminescent and bioluminescent techniques. Clin Chem. 1991;37(9):1472–81. pmid:1893571
5. Gingrich JC, Davis DR, Nguyen Q. Multiplex detection and quantitation of proteins on western blots using fluorescent probes. Biotechniques. 2000;29(3):636–42. pmid:10997278
6. Eaton SL, Hurtado ML, Oldknow KJ, Graham LC, Marchant TW, Gillingwater TH, et al. A guide to modern quantitative fluorescent western blotting with troubleshooting strategies. J Vis Exp. 2014(93):e52099. pmid:25490604
7. Taylor SC, Posch A. The design of a quantitative western blot experiment. Biomed Res Int. 2014;2014:361590. pmid:24738055
8. Kislinger T, Gramolini AO, Pan Y, Rahman K, MacLennan DH, Emili A. Proteome dynamics during C2C12 myoblast differentiation. Mol Cell Proteomics. 2005;4(7):887–901. pmid:15824125
9. Wagatsuma A, Shiozuka M, Kotake N, Takayuki K, Yusuke H, Mabuchi K, et al. Pharmacological inhibition of HSP90 activity negatively modulates myogenic differentiation and cell survival in C2C12 cells. Mol Cell Biochem. 2011;358(1–2):265–80. pmid:21739150

11

Total Protein Analysis as a Reliable Loading Control for Quantitative Fluorescent Western Blotting

Samantha L. Eaton[1] , Sarah L. Roche[2] , Maica Llavero Hurtado[1] , Karla J. Oldknow[3] , Colin Farquharson[3] , Thomas H. Gillingwater[2,4], Thomas M. Wishart[1,4]

[1] Division of Neurobiology, The Roslin Institute and Royal (Dick) School of Veterinary Studies, University of Edinburgh, Edinburgh, United Kingdom
[2] Centre for Integrative Physiology, University of Edinburgh, Edinburgh, United Kingdom
[3]Developmental Biology, The Roslin Institute and Royal (Dick) School of Veterinary Studies, University of Edinburgh, Edinburgh, United Kingdom
[4] Euan MacDonald Centre for Motor Neurone Disease Research, University of Edinburgh, Edinburgh, United Kingdom

ABSTRACT

Western blotting has been a key technique for determining the relative expression of proteins within complex biological samples since the first publications in 1979. Recent developments in sensitive fluorescent

Citation: Total Protein Analysis as a Reliable Loading Control for Quantitative Fluorescent Western Blotting. Eaton SL, Roche SL, Llavero Hurtado M, Oldknow KJ, Farquharson C, et al. (2013) PLoS ONE 8(8): e72457. doi:10.1371/journal.pone.0072457

labels, with truly quantifiable linear ranges and greater limits of detection, have allowed biologists to probe tissue specific pathways and processes with higher resolution than ever before. However, the application of quantitative Western blotting (QWB) to a range of healthy tissues and those from degenerative models has highlighted a problem with significant consequences for quantitative protein analysis: how can researchers conduct comparative expression analyses when many of the commonly used reference proteins (e.g. loading controls) are differentially expressed? Here we demonstrate that common controls, including actin and tubulin, are differentially expressed in tissues from a wide range of animal models of neurodegeneration. We highlight the prevalence of such alterations through examination of published "–omics" data, and demonstrate similar responses in sensitive QWB experiments. For example, QWB analysis of spinal cord from a murine model of Spinal Muscular Atrophy using an Odyssey scanner revealed that beta-actin expression was decreased by $19.3 \pm 2\%$ compared to healthy littermate controls. Thus, normalising QWB data to β-actin in these circumstances could result in 'skewing' of all data by $\sim 20\%$. We further demonstrate that differential expression of commonly used loading controls was not restricted to the nervous system, but was also detectable across multiple tissues, including bone, fat and internal organs. Moreover, expression of these "control" proteins was not consistent between different portions of the same tissue, highlighting the importance of careful and consistent tissue sampling for QWB experiments. Finally, having illustrated the problem of selecting appropriate single protein loading controls, we demonstrate that normalisation using total protein analysis on samples run in parallel with stains such as Coomassie blue provides a more robust approach.

BACKGROUND

Biochemical analysis using Western blotting is an essential tool in determining relative protein expression in complex biological samples. It is often used in conjunction with mass screening technologies such as proteomics to confirm differential candidate expression in various models of disease. Together with increasingly sophisticated *in vivo* and *in vitro* biological models, quantitative protein expression analyses are frequently being employed in attempts to elucidate the molecular mechanisms regulating cellular form and function in health and disease.

Traditionally, Western blotting with ECL (enhanced chemilluminescence) has been referred to as a semi-quantitative technique due to the lack of cumulative luminescence linearity and limited quantitative reproducibility [1]. With the development of more sensitive fluorescent labelling, which demonstrates a greater quantifiable linear range, sensitivity and stability in comparison to conventional ECL detection [2], analysis of protein expression can be justifiably termed "quantitative". It is therefore imperative to ensure uniformity of sample loading with an even greater degree of precision to avoid erroneous data acquisition when using these higher resolution tools [3]. The leading company in this market is LICOR and its Odyssey fluorescence imaging scanner appears to have the most significant market penetration at present with 206 instruments currently installed in the UK alone (Personal communication; LICOR technical consultants).

In order to accurately measure protein levels in a sample, "loading control" (LC) proteins are commonly used as internal standards. The loading controls are generally derived from ubiquitously expressed "housekeeping" genes and have been widely used due to their presumed consistent level of expression across a diverse range of samples. Actin and tubulin are two of most frequently used loading controls in biomedical research, however an increasing number of studies have suggested they may be differentially expressed in animal and experimental models [4]–[8]. Furthermore, LCs may also differ in expression from tissue to tissue or following exposure to infectious agents [9]. Therefore, normalising data according to loading control protein expression could further skew results leading to erroneous conclusions.

This study set out to specifically investigate the reliability of loading controls as internal standards, and characterise a robust, reproducible and simple method for normalising protein load when practicing modern quantitative Western blotting. We present data focusing on the expression levels of commonly used loading control proteins in the nervous system, as this is our specific area of research interest. However, we go on to demonstrate the importance of accurate loading controls for research on a broad range of biological tissues and demonstrate that normalisation using total protein analysis (TPA) with stains such as Coomassie and Instant blue provides a more robust baseline for performing QWB experiments.

RESULTS AND DISCUSSION

Expression Levels of Commonly used "Loading Control" Proteins

In order to obtain an initial estimate for the variability of expression levels of commonly used loading control proteins in experimental studies, we first undertook an examination of published protein expression data from a range of human and animal model studies including four of our own published datasets (Table 1; [6], [10]–[16]). With the increased stringency requirements for mass screening reporting, raw data sets are publicly available for many experimental comparisons, allowing further examination of the data for proteins of interest which may not have been highlighted in the full manuscript. We began with a focus on the expression levels of commonly used loading control proteins within the nervous system, as this is our specific area of research interest. We searched these datasets for expression data on a range of cytoskeletal proteins (actin, actinin and various tubulin isoforms), mitochondrial proteins (VDAC1 and VDAC2) and a nuclear protein (HC1), all of which are commonly used as a loading control or internal reference proteins for expression value normalisation in comparative protein quantitation experiments (Table 1).

Table 1. Examination of publicly available mass screening data reveals alterations in expression levels of commonly used "loading control" proteins.

Protein Name	Symbol	Protein ID	Localisation	Alteration	Model	Tissue	Technique	Reference
Actin	Actc1	IPI00654242.1	Cytoskeleton	Upregulated (3.07 fold)	SMA mouse	Cardiac Muscle	Label Free Proteomics	Mutsaers et al 2011 [6]
Actin	Acta1	IPI00110850.1	Cytoskeleton	Upregulated (3.01 fold)	SMA mouse	Skeletal Muscle	Label Free Proteomics	Mutsaers et al 2011 [6]
Actin (Unspecified Isoform)	-	p60710	Cytoskeleton	Absent	Wlds mouse	Striatal Synapses	2D Gel Proteomics	Wishart et al 2007 [10]
actinin, alpha 1	Actn1	Q7TPR4	Cytoskeleton	Downregulated (0.6 ratio)	SOD1 mouse	Cortical Synapses	iTRAQ Proteomics	Flynn et al 2012 [11]
Histone Cluster 1	HC1	NP_056601	Nuclear	Upregulated (1.3 ratio)	AOPE mouse	Sciatic Nerve	iTRAQ Proteomics	Comely et al 2011 [12]
Tubulin Alpha 1B Chain	ActA1B	IPI00117348.4	Cytoskeleton	Upregulated (1.32 fold)	SMA mouse	Skeletal Muscle	Label Free Proteomics	Mutsaers et al 2011 [6]

Tubulin Beta 4 Chain	TubB4	IPI00109073.5	Cytoskel-eton	Downregu-lated (−1.16 ratio)	SMA mouse	Hippo-campus	iTRAQ Proteomics	Wishart et al 2010 [13]
Tubulin beta-4 chain	Tubb4	N/A	Cytoskel-eton	Downregu-lated (−0.28 ratio)	Ercc1 mouse	Hippo-campal synapses	iTRAQ Proteomics	Vegh et al 2012 [14]
tubulin, alpha 4a	TubA4A	P68366	Cytoskel-eton	Downregu-lated (−1.13 fold)	FTLD patient	Cerebel-lum	Gel Separa-tion & Label free	de-Souza et al 2012 [15]
tubulin, beta 2a	Tubb2A	Q9CWF2	Cytoskel-eton	Upnregu-lated (0.26 ratio)	SOD1 mouse	Cortical Synapses	iTRAQ Proteomics	Flynn et al 2012 [11]
tubulin, beta 2a	Tubb2A	Q9CWF2	Cytoskel-eton	Upregulated (0.26 ratio)	FTLD patient	Prefrontal Cortex	Gel Separa-tion & Label free	de-Souza et al 2012 [15]
Voltage Depen-dant anion Channel 1	VDAC1	q60932	Mito-chondrial	Downregu-lated (−4.08 fold)	Wlds mouse	Striatal Synapses	2D Gel Proteomics	Wishart et al 2007 [10]
Voltage Depen-dant anion Channel 1	VDAC1	q60932	Mito-chondrial	Upregulated (96.4%)	PPT1 mouse	Cortex	Quantitative Western Blot	Kielar et al 2009 [16]
Voltage Depen-dant anion Channel 1	VDAC1	q60932	Mito-chondrial	Downregu-lated (−0.43 ratio)	SOD1 mouse	Cortical Synapses	iTRAQ Proteomics	Flynn et al 2012 [11]

Note: Alterations listed in table are in units published in the original manuscript or accompanying supplementary material.

This initial analysis revealed evidence for differential expression of all loading controls assessed using mass screening tools, including both -array technologies and proteomics. Discrepancies in expression of loading control proteins were detected across a diverse range of conditions, including Alzheimer's disease, lysosomal storage disorders and the motor neurone disease Spinal Muscular Atrophy (SMA). Notably, this re-interrogation of published data also identified differential expression of loading control proteins across an assortment of tissues sampled, an issue of potential critical importance for subsequent data normalisation and post-omics validation.

Actin and Tubulin are differentially Expressed in Pathologically-affected Tissue from a Mouse Model of SMA

Given that our analysis of raw data from published protein expression studies revealed widespread differential expression of common loading controls in various pathological conditions and neurological diseases ([10], [17], cf. Table 1), we next wanted to demonstrate that such alterations can also be detected when using modern QWB techniques. Using spinal cord tissue harvested from an established mouse model of severe SMA [18]–[19], we next quantified levels of β-actin and β-tubulin proteins. We found altered expression levels of both β-actin and β-tubulin when comparing the spinal cord of SMA affected mice (SMN:SMN2) with littermate controls (figure 1a. and 1b. respectively) using QWB. Both β-actin and β-tubulin expression were significantly down regulated in SMA compared to control tissue, by 19.3±2% and 7.3±0.5% respectively (mean ± SEM). Moreover, determination of total protein load demonstrated a high level of uniformity with a difference of 1.8±0.4% (mean ± SEM) between wild type and affected mice (n=6) (figure 1c). Thus, altered LC protein expression as highlighted by proteomic studies on tissues from neurodegenerative conditions (including SMA; see Table 1) can also be detected by quantitative Western blot.

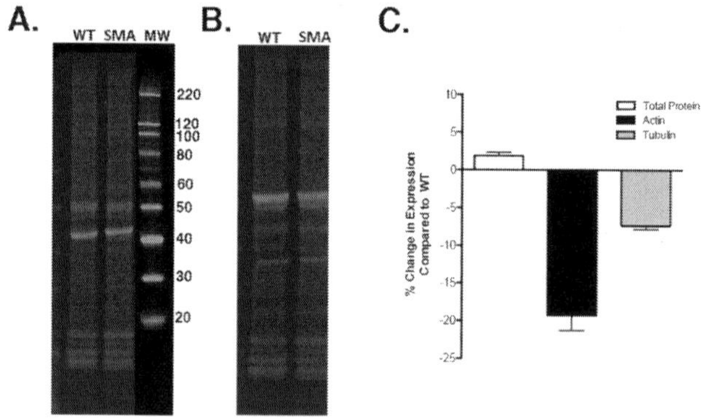

Figure 1. QWB for β-actin and β-tubulin demonstrates altered expression in pathological tissue.

QWB for the commonly used loading controls actin and tubulin in SMA spinal cord extracts. A. Overlay of β-Actin QWB in green (42 kDa) with total protein stained gel (red). B. Overlay of β-tubulin QWB in green (60 kDa) with total protein stained gel (red). C. Quantification of percentage change in the expression levels of β-actin (black bar) and β-tubulin (grey bar) when comparing SMA mice (SMA) to wild-type controls (WT). Total protein stain (white bar) is used as a control.

Differential β-actin Expression is Detectable by QWB across Multiple Tissues from Wild-type Mice

Given that significant alterations in common loading control proteins can be detected in affected tissues sampled from a range of disease models (Table 1), we next wanted to establish whether similar differences in expression could be identified in healthy ('wild-type') tissue, and also establish whether several tissue types could be differentially affected within the same individual. This latter issue is particularly pertinent for many neurodegenerative diseases (including SMA), where multi-system pathology is now being reported (e.g. [20]). QWB screening of multiple tissues for expression of candidate proteins is common practice and is an especially critical procedure when comparing systemic protein expression profiles, identifying biomarkers for disease progression in peripherally accessible tissues, or validating regulatory proteins during genetic or pharmacological manipulation. It is therefore essential for a loading control to exhibit stable expression across a wide array of tissues. However, our studies using C57Bl/6 ('wild-type') mice demonstrated that expression of β-actin varied considerably across different tissues from the same mouse (figure 2 A & B). In order to verify that the variability of β-actin expression was not due to loading error, quantification of the protein load of each tissue was carried out on a series of mass ranges corresponding to protein electrophoretic migration (figure 2C). These data demonstrated low individual variation of protein load across the different tissue samples within the series of molecular weight ranges measured. Variability ranged from only 1.87% (SEM) in the 40–80 kDa range up to 5.65% for the broadest mass range of 10–160 kDa across all tissues. Therefore, consistency of load across each of the tissues was demonstrated and was independent of the mass range measured (figure 2C).

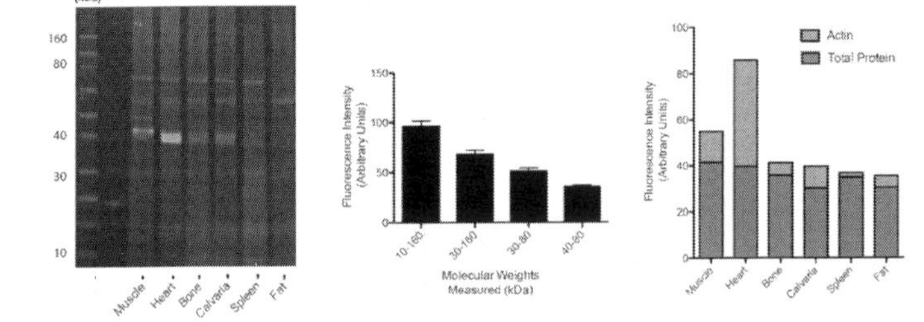

Figure 2. Comparative analysis of **β-actin** expression is highly variable across a broad range of tissues.

A. Representative images of the tissue samples in which actin expression was assessed. From left to right: Muscle (Gastrocnemius), heart, bone (femur), calvaria, spleen and fat (gonadal). Scale bar=1 cm. B. LICOR image of QWB demonstrating considerable variability of β-actin expression (green) in muscle, heart, bone, calvaria, spleen and fat extracts. Total protein stain gel image (red) is overlaid on QWB as a control. C. Total protein measurements for different molecular weight ranges demonstrates the accuracy of protein loading across the different tissue samples. D. Stacked bar graph demonstrating the comparative variability of β-actin (green bars) and total protein measurements (red bars) for each tissue examined.

Expression of β-actin was strikingly different when comparing a diverse range of tissues (figure 2B & D). An appreciable disparity in β-actin expression was observed between heart and spleen tissue, with a difference of 44 arbitrary fluorescence units (AFU) contrasting with the consistent total protein load demonstrated in the 40–80 kDa molecular weight range (figure 2D). Critically, normalising data to β-actin expression for cross tissue comparisons could therefore result in skewing of data by up to 22 fold when comparing these tissues. However, there was some homology in β-actin expression between certain tissues such as bone and fat (5.33 AFU and 4.99 AFU respectively), and for these tissues it may therefore be acceptable to use β-actin as an internal control, but only when comparing these specific samples. Interestingly, a similar issue regarding β-actin has been raised with cross-tissue RNA expression profiling. For example, a qRT-PCR study in fish concurs with our findings suggesting that β-actin is not an ideal loading control for certain tissue comparisons particularly for heart, muscle and brain, whereas its expression is more consistent between kidney and spleen [9]. Our results therefore demonstrate that it would be advisable to use total protein expression as an indication of protein load to accurately perform a comparative expression analysis across a wide range of tissue samples.

Loading Control Expression is not Homogeneous throughout Structurally Asymmetrical Tissues

Having demonstrated that expression of common loading controls such as actin was not necessarily consistent when comparing different tissues from the same animal, we next wanted to establish whether or not uniformity of protein expression was preserved throughout a single tissue from the same animal. Anecdotally it appears that researchers assume that loading control protein expression is stable within a given organ or tissue, with methodology sections of manuscripts routinely detailing gross, rather than specific, anatomical terms to describe tissue harvesting. To assess consistency within a biologically-relevant tissue, we measured and compared β-actin and neurofilament-light (NF-L) levels in proximal versus distal portions of the same mouse sciatic nerve using QWB. We found that levels of these LC proteins were not consistent throughout the two portions of the same nerve (Figure 3). Once again, we found that total protein load was consistent when quantified, however β-actin (a predominantly cytoplasmic isoform; [21] labelling was significantly higher ($p=0.0045$), 52% greater ±14.3% (SEM), when comparing the proximal to distal portion of the sciatic nerve (figure 3C & E). In contrast expression of neurofilament (NF-L), a major component of the neuronal cytoskeleton [22], was appreciably higher, nearly 8 fold with up to a 4 fold SEM in the distal portion of the sciatic nerve (figure 3D & F). Therefore, these results stress that accuracy and consistency of dissection are crucial, even when evaluating the same tissues from a single animal for comparative analysis. Whilst this should be standard practice regardless, it is especially true for structurally asymmetrical tissues and failure to do so could have significant consequences for both -omics screens (such as proteomic comparisons) and for QWB analyses.

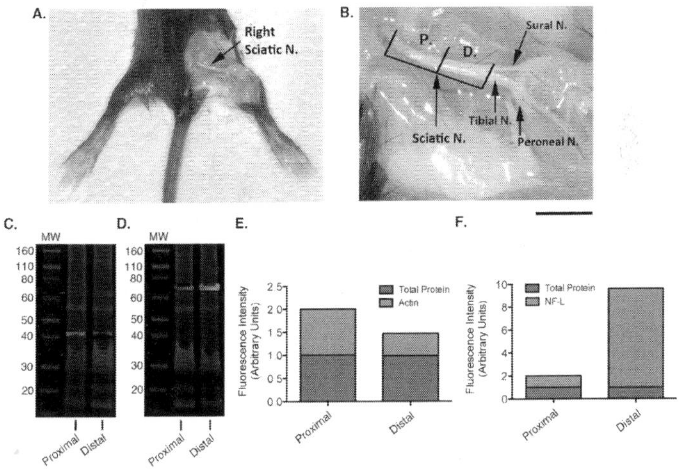

Figure 3. Actin & NF-L levels are not stable throughout different regions of the mouse sciatic nerve.

Differential expression in proximal versus distal sciatic nerve preparations. A. Photograph depicting the lower half of a Bl6 mouse with sciatic nerve exposed on the right leg. B. Higher magnification photograph shows sciatic nerve and subsequent branches (anatomical nomenclature taken from [26]). Scale bar: A=1 cm, B=0.5 cm. C. Representative LICOR overlay image of β-actin QWB (green) and total protein stained gel (red) in proximal and distal portions of sciatic nerve from the same mouse. D. Representative LICOR overlay image of NF-L QWB (green) and total protein stained gel (red) in proximal and distal portions of sciatic nerve from the same mouse. E. Stacked bar graph demonstrating the comparative expression of β-actin (green bars) and total protein stain (red bars) expression in proximal and distal sections of sciatic nerve. F. Comparative expression of NF-L (green bars) and total protein stain (red bars) expression in proximal and distal sections of sciatic nerve.

Limitations in Standard Loading Controls: β-actin and β-tubulin Working Range and Sensitivity Explored

The ideal internal loading control protein for QWB must be abundant with a wide linear range of detection to accommodate proteins of varying levels of expression. In order to evaluate the suitability of β-actin and β-tubulin as loading controls for QWB we tested both their working linear range and sensitivity by quantifying their expression throughout a dilution series, ranging from 1 to 40 μg of protein, produced from mouse whole brain tissue homogenate (figure 4a & b). The working range of β-actin where linearity was maintained was between 1 and 30 μg of protein loaded (figure 4a & b). Here, our quantitative Western blotting data conflicts with that of others who have reported that the linear range of β-actin was far smaller, only up to 2 μg of protein before saturation of the signal occurred [4], [8]. However these studies used less sensitive ECL detection methodologies therefore this disparity is most probably caused by the limitation of ECL based imaging and "quantification".

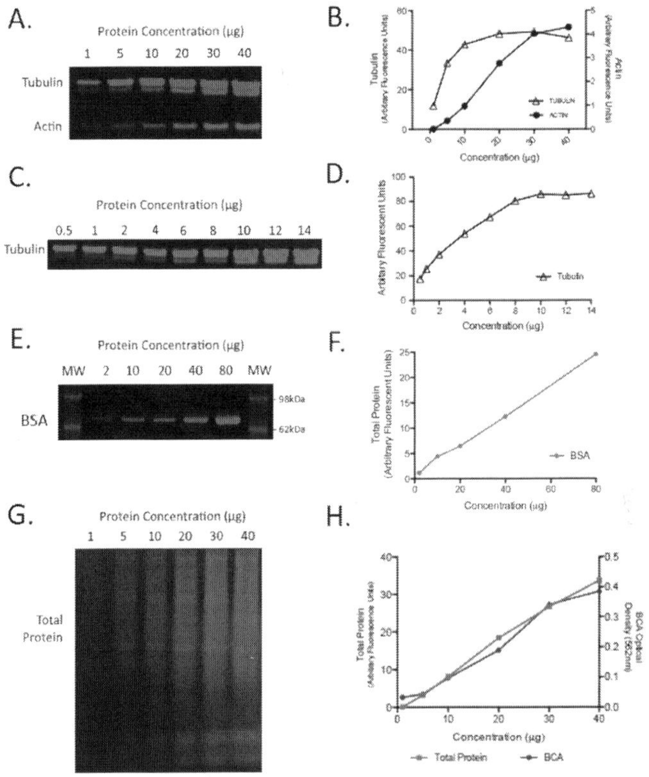

Figure 4. Linear range and sensitivity of total protein stain is greater than the conventional loading controls **β-actin** or **β-tubulin.**

(A) Representative LICOR image for a protein dilution series of whole brain homogenate 1, 5, 10, 20, 30, 40 μg demonstrating the working range of β-actin and β-tubulin when using QWB. (B) Quantification of protein dilution series showing the linear ranges of β-actin (black circle) and β-tubulin (open triangle). Note that tubulin expression appears to saturate at less than 10 ug of brain homogenate. (C) In order to pinpoint the saturation level a tubulin specific protein dilution series over a smaller range (0.5, 1, 2, 4, 6, 8, 10,12 and 14 μg) establishing the saturation point of β-tubulin when using QWB. (D) Quantification of β-tubulin linear range. (E) Total protein stain of dilution series 2, 10, 20, 40, 80 μg made using the bovine serum albumin standard (2 μg/μl) from the Pierce BCA kit (see methods). BSA molecular weight is 66.5 kDa. Imaging of this dilution series demonstrates imaging of a broad concentration range without saturation at a single protein mass. (F) Graphical representation of quantification from BSA dilution series in panel E. This demonstrates wide linear detection and high correlation

(0.998) validating the use of total protein measurements as a viable method for detecting protein load using the LICOR system. (G) Total protein stain of whole brain homogenate dilution series 1, 5, 10, 20, 30, 40 μg demonstrates the broad concentration range detectable without saturation. (H) Correlation between the total protein stain quantification (red line) and the BCA OD (blue line) for the protein dilution series demonstrates wide linear detection and high correlation (0.996 & 0.979 respectively) validating the use of total protein measurements as a viable "loading control" for QWB using the LICOR system.

When examining β-tubulin, it was evident that this protein was so abundant in brain extracts that the signal began to saturate out at less than 10 μg of protein (figure 4 a & b). In order to accurately determine the precise linear range of β-tubulin a tighter dilution series of protein load, 0.5 to 14 μg, was employed (figure 4c and d). Saturation of the β-tubulin signal occurred at 8–10 μg of protein load. Again, our QWB results conflicted with previous ECL studies suggesting that the linear range of β-tubulin peaks at 5 μg [4] and it is assumed that this is also likely due to the limitation of ECL based imaging and "quantification". Our results suggest that β-tubulin rather than β-actin would be a more appropriate loading control when detecting low-abundance proteins in homogeneous extracts, however actin has a greater working range of sensitivity. In addition, our data has emphasised the superior sensitivity of fluorescent QWB in comparison to ECL as we demonstrate a far wider working range for both of these commonly used single protein loading controls.

Total Protein Analysis is an Accurate Measure of Protein Load

Our analyses detailed above have highlighted several important variables that need considering when choosing an internal control for QWB, including but not limited to the linear range of sensitivity and disparities in expression across different tissue samples or within portions of the same tissue. Finally, we wanted to establish whether a total protein analysis approach would provide a more reliable and accurate measure of protein load for QWB experiments. To begin, we assessed the detection sensitivity using a dilution series created using a bovine serum albumin (BSA) protein standard (Figure 4E & F). BSA has a single band at a molecular weight of 66.5 kDa. We were able to detect a linear fluorescence profile across a broad concentration range as determined by the coefficient of variation (R^2 value) of 0.998. Importantly, as the BSA standard dilution series effectively means that all of the protein loaded is represented by a single band, this not only validates the linear nature of detection using this system but also demonstrates the lack of saturation typically found with ECL based systems.

Finally, when applied to real biological samples using the LICOR Odyssey quantitative scanning system, total protein analysis (using coomassie) was linear in its detection across a broader range of protein loading than either actin or tubulin (1 to 40 ug; Figure 4G & H). The coefficient of variations for both Bicinchoninic Acid solution (BCA) assay and total protein analysis were 0.979 and 0.996 respectively. These R^2 values demonstrated a high degree of linearity in both assays. Moreover, total protein analysis correlated directly with BCA protein concentration data (Figure 4G & H) throughout the broad 1–40 μg protein load dilution series further indicating its reliability as a control for protein load using the Odyssey quantitative imaging system. We therefore suggest that the use of total protein analysis provides a measure of protein load that circumvents many of the problems associated with the use of single loading control proteins: it is unchanged when comparing tissues from different models (c.f. Figure 1); it is consistent across different tissue types (c.f. Figure 2); and different portions of the same tissue (c.f. Figure 3).

CONCLUSIONS

Western blotting has traditionally been a "semi-quantitative" technique using house keeping genes as internal reference standards. These standards are required to compensate for any technical errors that may have arisen due to issues such as poor transfer or unequivocal loading. However, our QWB studies have demonstrated a critical problem with the use of some common loading controls for this role. Differential expression of commonly used loading control proteins including β-actin and β-tubulin, occurs when comparing a wide range of tissues, when examining different portions of the same tissue and when pathological conditions arise. Therefore if normalisation (using single protein loading control expression as a correction factor) of quantitative Western blotting results is required, all of the resulting data could be skewed as a result of the differential expression of a single protein [23].

Total protein analysis is an alternative simple technique in QWB to accurately determine if equivalent protein loading has been achieved within a gel [24]. Data obtained by total protein analysis is independent of the pitfalls that can occur using "common" house keeping genes as loading controls. That is not to say that housekeeping genes cannot be used as loading controls, but that they should only be used in a limited fashion once the researcher has fully investigated sample expression homogeneity for the gene in question, and if the protein load falls within the working range of that particular loading control. Consequently, we propose it would be prudent to use total protein analysis to save time, resources, increase sensitivity and accuracy as well as the working range of protein load for quantitative Western blotting. Total protein analysis should therefore be considered an alternative standard reference for data normalisation in modern quantitative fluorescent Western blotting.

MATERIALS AND METHODS

Tissue Harvesting and Protein Extraction

Ethics statement

All animal experiments were approved by a University of Edinburgh internal ethics committee and were performed under license by the UK Home Office (project license number 60/3891).

Preparation of severe model of Spinal Muscular Atrophy (SMA) spinal cord homogenates

Spinal cord were harvested from SMN/SMN2 severe model of SMA and wild-type controls at P5 and homogenised in RIPA buffer (Sigma, UK) containing 5% protease inhibitor cocktail (Roche) as previously described [13]. All SMA based comparisons were therefore Smn−/−;SMN2 (SMA) compared to wild-type (WT) litter mate controls unless otherwise stated. Protein was extracted and concentrations determined using a BCA assay (Pierce) according to manufacturers instructions, as previously described [10], [13], [25].

Preparation of range of tissue samples from C57/black mice

Quadraceps femoris muscle, gonadal fat, heart, calvaria, spleen and tibial bone tissue were dissected from 10 day old C57Bl/6 mice. Tissues were processed as outlined above.

Preparation of sciatic nerve tissue

Proximal and distal sections of the sciatic nerve from *C57/black mice* were dissected in 1xPBS (phosphate buffered saline) and frozen immediately on dry ice. The samples were homogenised in iTRAQ buffer (6 M urea, 2 M thiourea, 2% CHAPS, 0.5% SDS), sonicated 5×10 secs on ice, and centrifuged at 14 K for 30 minutes. 1:100 dilutions of supernatant: dH_20 were used in a BCA assay following manufacturers instructions.

Quantitative Western Blotting

Samples were denatured in NuPage® LDS Sample buffer 4X (Invitrogen, UK) at 98°C and 15 ug of protein loaded (with the exception of the protein dilution series) and run on commercially produced pre-cast 4–12% Bis-Tris gels (Invitrogen). Gels were run in duplicate in parallel in the same electrophoretic tank at the same time. One gel was stained using Instant blue (Expedeon) or coomassie (see total protein stain below) and one was used to transfer the protein to a polyvinylidene fluoride (PVDF) membrane using the I-Blot® transfer system (Invitrogen, UK) using programme 3 for 8.5 minutes. Membranes were incubated with Odyssey blocking buffer (Li-Cor) prior to incubation with rabbit polyclonal antibodies directed against β-actin (1:1000, Abcam 8226), β-tubulin (1:1000, Abcam 8226) and mouse monoclonal anti-NF-L (Millipore AB9568) overnight at 4°C. Goat anti-rabbit IgG (H+L) 800 CW, goat anti-rabbit (680 RD) and/or goat anti-mouse (H+L) was applied for 90 minutes at room temperature (1:5000, LI-COR) prior to washing with PBS. Visualisation and quantification was carried out with the LI-COR Odyssey® scanner and software (LI-COR Biosciences). Blots (and gels) were imaged using an Odyssey Infrared Imaging System Scan resolution of the instrument ranges from 21 to 339 μm, and in this study blots (and gels) were imaged at 169 μm. Quantification was performed on single channels with the analysis software provided as previously described [10], [13], [25].

Total Protein Gel Stain

All total protein stains within this manuscript have been carried out on gels and not membranes unless otherwise stated. As such there are caveats which should be taken into consideration if comparing total protein stained gels with membranes due to variables accompanying membrane transfer which are not accounted for when using this approach (see below). Post electrophoresis gels (see above) were stained using either Instant Blue (Expedeon) or

Coomassie (0.1% Coomassie R250, 40% methanol, 10% acetic acid) solution. Gels were left in Instant blue for 1 hour and washed in dH_20 prior to visualisation. Coomassie stained gels were left in Coomassie solution for 1 hour and de-stained using several washes in de-stain solution (40% methanol, 10% acetic acid) and then washed in dH_20 prior to visualisation. Stained gels were imaged directly on the Li-COR Odyssey® scanner using the 700 channel and quantified using the Odyssey® software. See above. Incidentally the instant blue stain can be visualised in both the 700 and 800 channels however it has greater resolution in the 700 channel. For total protein stains to be of use as loading controls, ideally the membranes to be probed should be stained directly for load. However, there are limitations with this in modern fluorescent imaging systems. Coomassie is not as effective on most PVDF membrane types when compared to stained gels as the background auto-fluorescence is naturally higher. If used directly on a membrane coomassie does not strip to allow for re-probing in the same imaging channel. Ponceau stains are reversible but staining is difficult to visualise and remains to be proven linear in its adherence. Moreover, any stain which is not blue in presentation or does not fluoresce in the 680 or 800 wavelength channels used by the LICOR Odyssey scanner can not be imaged using this system and measurement scaling will therefore differ from the quantitative fluorescent blots for candidates of interest. Parallel commercially produced precast gels should therefore be used to reduce polyacrylamide matrix variability; they should be loaded at the same time using the same "master mix" (i.e. protein/water/loading dye) in the same tank (multi gel capacity) in order to be run from the same powerpack under the same conditions. A further variable when comparing a probed membrane to a stained gel are alterations introduced by variable transfer efficiency. Potential inconsistency in transfer efficiency generally occurs according to differential molecular weight rather than inter lane variability i.e. higher efficiency of transfer with lower molecular weight. The use of commercially procured transfer packs (such as the I-Blot® stacks; see above) coupled with a rapid semi-dry fast transfer system as detailed above should further improve reproducibility. By being aware of the steps where possible error could be introduced and taking the appropriate precautions such as those listed here, inter gel variability and transfer variance should be kept to a minimum, and appropriate data interpretation can be expected within the limitations of the system.

Calculation of Linear Ranges of β-actin and β-tubulin

Concentration of protein extracts can be determined in a variety of ways. The most commonly used may be the Bicinchoninic Acid assays (BCA). BCA assays involve reduction of copper ions in a temperature dependant fashion with the level of reduction correlating with protein concentration. Reduced copper ions bind to BCA forming a purple product which can be detected at 562 nm. Each run includes a dilution series of a known protein standard – bovine serum albumin (BSA) as a reference curve to allow determination of absolute protein concentrations. As each reaction set can be subtly influenced by incubation time and temperature, samples which will be grouped together for analysis should routinely be assayed together against the same standard curve. Here we employed a series of protein dilutions (1, 5, 10, 20, 30, 40 μg) and (0.5, 1, 2, 4, 6, 8, 10, 12, 14 μg) which were produced from mouse whole brain homogenate. Preparation of the dilution series was carried out after the concentration of protein had been determined using a micro BCA assay (Pierce). Briefly, two 4–12% Bis-tris gels were loaded; one stained for total protein (Coomassie or Instant Blue) and the other was transferred for QWB as above. Visualisation and quantification was carried out using the Li-COR Odyssey imager and software. See above.

Photography

Photographs of sciatic nerve dissection and organs were taken using a Nikon D200 camera with 105 mm micro NIKKIOR F2.8 lens.

Data Analysis and Figure Production

QWB data was analysed using Odyssey software as per manufactures guidelines and as previously described [10]. Data was graphed and statistical comparisons carried out using GraphPad Prizm as previously described [25]. Image overlays were produced using Adobe photoshop to overlay 700 & 800 channels obtained from the Odyssey imager (LICOR Biosciences).

ACKNOWLEDGMENTS

The authors would like to thank Norrie Russell of the Roslin Institute for his assistance with photography.

FUNDING STATEMENT

Work in the authors' laboratories has been supported by grant funding from the Biotechnology and Biological Sciences Research Council (BBSRC) UK through a studentship award (CJO) and Institute Strategic Programme Grant Funding from the BBSRC (CF & TMW) and the SMA Trust and Muscular Dystrophy Campaign (TG & SR). The funders had no role in study design, data collection and analysis, decision to publish, or preparation of the manuscript.

REFERENCES

1. Zellner M, Babeluk R, Diestinger M, Pirchegger P, Skeledzic S, et al. (2008) Fluorescence-based Western blotting for quantitation of protein biomarkers in clinical samples. Electrophoresis 29: 3621–3627.
2. Gingrich JC, Davis DR, Nguyen Q (2000) Multiplex detection and quantitation of proteins on western blots using fluorescent probes. Biotechniques 29: 636–42.
3. Aoki K, Yamada M, Kunida K, Yasuda S, Matsuda M (2011) Processive phosphorylation of ERK MAP kinase in mammalian cells. Proc Natl Acad Sci USA 108: 12675–12680.
4. Suzuki O, Koura M, Noguchi Y, Uchio-Yamada K, Matsuda J (2011) Use of sample mixtures for standard curve creation in quantitative Western blots. Exp Anim 60: 193–196.
5. Wishart TM, Pemberton HN, James SR, McCabe CJ, Gillingwater TH (2008) Modified cell cycle status in a mouse model of altered neuronal vulnerability (slow Wallerian degeneration (Wlds). Genome Biol 9: R101 doi:10.1186/gb-2008-9-6-r101
6. Mutsaers CA, Wishart TM, Lamont DJ, Riessland M, Schreml J, et al. (2011) Reversible molecular pathology of skeletal muscle in spinal muscular atrophy. Hum Mol Genet 20: 4334–4344.
7. Castaño Z, Kypta RM (2008) Housekeeping proteins: Limitations as references during neuronal differentiation. The open access Neuroscience Journal 2: 36–40.
8. Dittmer A, Dittmer J (2006) β-Actin is not a suitable loading control in Western blot analysis. Electrophoresis 27: 2844–2845.
9. Dang W, Sun L (2011) Determination of internal controls for quantitative real time RT-PCR analysis of the effect of Edwardsiella tarda infection on gene expression in turbot (Scophthalmus maximus). Fish & Shellfish Immunol 30: 720–728.
10. Wishart TM, Paterson JM, Short DM, Meredith S, Robertson KA, et al. (2007) Differential proteomics analysis of synaptic proteins identifies potential cellular targets and protein mediators of synaptic neuroprotection conferred by the slow Wallerian degeneration (Wlds) gene. Mol Cell Proteomics 6: 1318–1330.
11. Flynn JM, Czerwieniec GA, Choi SW, Day U, Gibson BW, et al. (2012) Proteogenomics of synaptosomal mitochondrila oxidative stress. Free Radic Biol Med 53: 1048–1060.

12. Comley LH, Fuller HR, Wishart TM, Mutsaers CA, Thomson D, et al. (2011) ApoE isoform-specific regulation of regeneration in the peripheral nervous system. Hum Mol genet 20: 2406–2421.
13. Wishart TM, Huang JP, Murray LM, Lamont DJ, Mutsaers CA, et al. (2010) SMN deficiency disrupts brain development in a mouse model of sever spinal muscular atrophy. Hum Mol Genet 19: 4216–4228.
14. Végh MJ, de Waard MC, van der Pluijm I, Ridwan Y, Sassen MJ, et al. (2012) Synaptic proteome changes in a DNA repair deficient ercc1 mouse model of accelerated aging. J Proteome Res 1: 1855–1867.
15. Martins-de-Souza D, Guest PC, Mann DM, Roeber S, Rahmoune H, et al. (2012) Proteomic analysis identifies dysfunction in cellular transport, energy, and protein metabolism in different brain regions of atypical frontotemporal lobar degeneration. J Proteome Res 11: 2533–2543.
16. Kielar C, Wishart TM, Palmer A, Dihanich S, Wong AM, et al. (2009) Molecular correlates of axonal and synaptic pathology in mouse models of Batten disease. Hum Mol Genet 18: 4066–4080.
17. Gebhardt FM, Scott HA, Dodd PR (2010) Housekeepers for accurate transcript expression analysis in Alzheimer's disease autopsy brain tissue. Alzheimer's & Dementia 6: 465–474.
18. Monani UR, Sendtner M, Coovert DD, Parsons DW, Andreassi C, et al. (2000) The human centromeric survival motor neuron gene (SMN2) rescues embryonic lethality in Smn(−/−) mice and results in a mouse with spinal muscular atrophy. Hum Mol Genet 9: 333–339.
19. Murray LM, Comley LH, Thomson D, Parkinson N, Talbot K, et al. (2008) Selective vulnerability of motor neurons and dissociation of pre- and post-synaptic pathology at the neuromuscular junction in mouse models of spinal muscular atrophy. Hum Mol Genet 17: 949–962.
20. Hamilton G, Gillingwater TH (2013) Spinal muscular atrophy: going beyond the motor neuron. Trends Mol Med 19: 40–50.
21. Dugina V, Zwaenepoel I, Gabbiani G, Clement S, Chaponnier C (2009) Beta and gamma-cytoplasmic actins display distinct distribution and functional diversity. J Cell Sci 122: 2980–2988.
22. Hirokawa N, Glicksman MA, Willard MB (1984) Organization of mammalian neurofilament polypeptides within the neuronal cytoskeleton. J Cell Biol 98: 1523–1536.

23. Aldridge GM, Podrebarac DM, Greenough WT. Weiler IJ (2008) The use of total protein stains as loading controls: an alternative to high-abundance single protein controls in semi-quantitative immunoblotting. J Neurosci Methods 172: 250–254.

24. Welinder C, Ekbald L (2011) Coomassie as loading control in Western blot analysis. J Proteome Res 10: 1416–1419.

25. Wishart TM, Rooney TM, Lamont DJ, Wright AK, Morton AJ, et al. (2012) Combining comparative proteomics and molecular genetics uncovers regulators of synaptic and axonal stability and degeneration in vivo. PLoS Genet 8: e1002936 doi:10.1371/journal.pgen.1002936

26. Silva DN, Coelho J, Frazílio Fde O, Odashiro AN, Carvalho Pde T, et al. (2010) End-to-side nerve repair using fibrin glue in rats. Acta Cir Bras 25: 158–162.

INDEX

S

T

U

V

W